WITHDRAWN

DEVELOPING THE FUTURE AVIATION SYSTEM

To Janice

Developing the Future Aviation System

Edited by
ROD BALDWIN
Managing Director
Baldwin International Services

Ashgate
Aldershot • Brookfield USA • Singapore • Sydney

© Rod Baldwin 1998

All rights reserved. No part of this publication may be reproduced, stored in a retrieval system, or transmitted in any form or by any means, electronic, mechanical, photocopying, recording or otherwise without the prior permission of the publisher.

Published by
Ashgate Publishing Ltd
Gower House
Croft Road
Aldershot
Hants GU11 3HR
England

Ashgate Publishing Company
Old Post Road
Brookfield
Vermont 05036
USA

British Library Cataloguing in Publication Data
Developing the future aviation system
  1. Aeronautics
 I. Baldwin, Rod
  387.7

Library of Congress Cataloging-in-Publication Data
Developing the future aviation system / edited by Rod Baldwin
     p. cm.
  ISBN 0-291-39843-X (hc.)
   1. Aeronautics -- Technological innovations. I. Baldwin, Rod
 TL500.D48 1998
  629.13--dc21   98-28581
                                                                                    CIP

ISBN 0 291 39843 X

Printed and bound by MPG Books Ltd, Bodmin, Cornwall

Contents

List of figures vii
List of tables viii
Preface ix

Part one: New concepts for aircraft and airports 1

1 A systems approach to developing the new aircraft 3
   Scott Jackson

2 New generation airports 20
   Jenny Beechener

3 The airport business and information technology 27
   Mark Sawle-Thomas

4 Airport security 39
   Peter Wilkins

Part two: Human factors and training 53

5 Human factors in the cockpit 55
   Margaret T. Shaffer

6 Laws for the design of the Universal Cockpit displays 66
   Lawrence E. Tannas Jr.

7 Creating a culture of safety 72
   Dianne Hill

8	Human factors in Air Traffic Control *V. David Hopkin*	85
9	Training issues in Air Traffic Flow Management *John A. Wise, V. David Hopkin and Daniel J. Garland*	110

Part three: Managing the aviation system 133

10	The Air Traffic Management System –present and future *Vince Galotti*	135
11	Improving capacity – implementation of the FANS CNS/ATM system in the Asia/Pacific region *Brian O'Keeffe*	148
12	The new IATA International Passenger Liability Regime *Lorne S. Clark*	177
13	Developments in aircraft interior design *Carole Favart-Andrieux*	187

List of figures

Figure 1.1	Aircraft system architecture (and ATA chapter correlation)	5
Figure 1.2	Matrix of aircraft operational functions	7
Figure 1.3	The perform flight operations function	8
Figure 1.4	Allocation of direct operating costs	14
Figure 1.5	DOC regimes	15
Figure 3.1	Business effects of an airport	28
Figure 3.2	Current management position of airports across the world	29
Figure 3.3	Proportion of non-aeronautical revenue at airports	31
Figure 5.1	Ground-Air Paradigm (GAP)	56
Figure 7.1	The R1234® concept - four Rs of leadership	73
Figure 7.2	The leadership grid®	76
Figure 7.3	Change by design	81
Figure 7.4	Aligned for safety or error	82
Figure 9.1	TFM system success	111
Figure 9.2	Aircraft situation display (ASD)	117
Figure 9.3	Dynamic density display	119
Figure 9.4	Capacity\ratio graph	120
Figure 9.5	Route density display	121

List of tables

Table 1.1.	Comparison of traditional and aircraft life-cycle functions	6
Table 1.2	Requirements affecting the aircraft system	12
Table 1.3	Wing sizing requirements	13
Table 1.4	Compatibility of the SE and certification processes	16
Table 2.1	Major new airport developments in China (PRC)	25
Table 2.2	Timetables for privatisation	26
Table 7.1	Decision-making criteria: effective use of human resources	74
Table 11.1	Geographical areas based on major traffic flows	158
Table 11.2	Explanation of terms used in tables	161
Table 11.3	Generic table for enroute operations	163
Table 11.4	Area: Australia/New Zealand-North America (via South Pacific)	168
Table 11.5	Area: Asia -North America (via North Pacific and the Russian Far East)	172
Table 12.1	List of carriers signatory to the IATA intercarrier agreement on passenger liability as at 24 January 1998	185

Preface

Major changes are taking place in the realm of technology and the effects are being seen in the world of aviation. In an industry which started with the concept of 'open skies' each sector has traditionally developed on its own and adjusted to developments in other areas as and when required.

Such unrelated developments have gradually reduced over the years but the need for integration is particularly important now as the skies become crowded and so many more people wish to travel. With increased commercialisation in many areas, there is also an increase interlocking between technological developments and the size of the financial investments required. Consequently, the aviation system has to develop as an integrated whole with all players having an awareness of what is happening in others sectors of the industry.

This book is intended to meet this requirement by addressing the breadth and depth of the aviation system and looking at some areas where significant developments are happening. In conjunction with following the processes of the developments the reader will see where the results might lead to in the next century. To support the integrated concept some overlap between chapters has been allowed where there is constructive inter-linkage of information.

As mentioned above, all of us in the aviation industry need to be aware of the whole picture and so this book should appeal to all persons who are moving into management functions and thus having to take account of matters which surround their particular responsibilities. By association, it will serve the same purpose in training centres, colleges institutions and universities which must make sure that students and trainees have access to information on matters which will shape their career development. Finally, there are the great numbers of people who just like 'being in aviation', either by employment or general interest.

To all these users, maintainers, operators, students and managers of the aviation system I hope that this book will be helpful and useful in supporting your activities.

Although the book is a broad selection of articles, for ease of reference it is divided into three broad themes. Part one deals with new concepts for aircraft and airports. Chapter one introduces a systems approach to designing aircraft, by Scott Jackson, which gives a guide to how this subject is developing at present and in the future. This is followed by Jenny Beechener who looks at the growth prospects of airports, the impact of market forces and how they are affected by organizational and regulatory matters. The contribution of information technology to shaping the changing nature of airports is explored by Mark Sawle-Thomas who introduces the concept of IT as a revenue generator rather than a costly overhead. Finally Peter Wilkins addresses the all important issue of the need for security against convenience and freedom of movement.

As so many aviation accidents are attributed to the 'Human Element', Part two covers a number of areas where great efforts are being made to gain a better understanding of the humans' role in the aviation system and how it can be improved. Chapter five begins with Human Factors in the cockpit where Margaret Shaffer covers a range of issues and makes the case for an integrated approach to cockpit design. This theme is continued by Lawrence Tannas who takes a more specific approach by enunciating a number of laws which should be followed in order to produce universal cockpit displays in the interests of safety. Creating a culture of safety through crew resource management is the theme by Dianne Hill who describes the application of a leadership grid to actually train flight crew in how to improve the way they work together. The role of Human Factors in Air Traffic Control are fully covered by David Hopkin with pertinent reviews of areas which need attention as levels of automation are increased and the role of controllers undergo change. This theme is continued by him in conjunction with John Wise and Daniel Garland in exploring the strategic environment of air traffic management and the need to train personnel for the different operating levels involved in the air traffic system.

Part three is concerned with air traffic management systems and managing the aviation system. Vince Galotti starts in Chapter ten by reviewing the ATM system and how a number of developments are meeting the need for a global integrated system which operates from gate-to-gate - especially the European ATM 2000+ strategy. How one such development was actually achieved is described by Brian O'Keeffe, in chapter eleven, with respect to the ICAO Future Air Navigation System (FANS). He also shows how FANS is needed and actually applied in the Asia\Pacific area. As change is affecting all institutions, it is important to remember that besides the need for technical excellence in the aviation system, there is also a need for high performance support from administrations and institutions. In chapter twelve Lorne Clark describes the procedures involved in producing a new international agreement on passenger compensation -

involving national civil aviation authorities, airlines, the Commission of the European Union, ICAO and IATA. Above all, the passenger wants to know what developments are happening in the cabin and here Carole Favart-Andrieux explains a number of aspects affecting the interior design, what the passenger might be looking for and what might be available in the future.

A word about the spelling and grammar used. An effort has been made to use British English throughout the book and all the authors have made great efforts to keep to this. However, with such a wide spread of authors there will occasionally be some differences which have crept in. For this the Editor accepts responsibility and apologizes. However, it should be noted that quotations and references are in their original language.

Finally, a big thank you to all the contributors who made this book possible. Their efforts are very much appreciated. Also, many thanks to my wife, Janice, who helped me put the book together.

Rod Baldwin
Luxembourg
August 1998.

The editor

Dr. Rod Baldwin is at present the Managing Director of Baldwin International Services (BIS) based in Luxembourg. Previously he was the Director of the Eurocontrol Institute of Air Navigation Services in Luxembourg after several years as the Principal of IAL Bailbrook College in Bath, England. This followed a distinguished academic career as a Research and Development engineer in the fields of electronics and communication engineering.

Part 1

NEW CONCEPTS FOR AIRCRAFT AND AIRPORTS

1 A systems approach to developing the new aircraft

Scott Jackson

Although born in the defence industry, the science of systems engineering (SE) has reached a level of maturity in which it is increasingly being applied in the commercial aircraft industry. For example, Petersen and Sutcliffe (1992) discuss the principles of SE as applied to aircraft development. In addition, major professional societies have technical committees and other activities devoted to the study of the SE process. These societies include the American Institute of Aeronautics and Astronautics (AIAA) and the Institute of Electrical and Electronic Engineers (IEEE). The Society of Automotive Engineering (SAE) has developed in cooperation with the Federal Aviation Administration (FAA) a set of certification guidelines, ARP 4754 (1996), which incorporate SE principles. The International Council on Systems Engineering (INCOSE) is devoted entirely to the development and application of the SE process.

SE is a process in which a complete system (for example, the global aviation system or a complete aircraft) is viewed as a whole and not as a collection of parts. The system is viewed as a hierarchy of functions from which requirements and solutions are derived. This chapter will concentrate on an aircraft system, consisting of the aircraft itself, its support and training elements, associated facilities, and personnel. The premise is that SE will both enable the development of aircraft which will fit into the larger global system and also facilitate the introduction of new technologies for that system.

SE also encompasses other processes, many of which predate the forulation of the SE concept. These processes include interface, risk, and configuration management. The SE concept of verification goes beyond testing (laboratory and flight) to include demonstration, analysis, and inspecion. SE requires that all requirements be verified.

The application of SE to commercial aircraft presents a set of requirements and processes unique to the commercial aircraft industry. SE can be applied to new, derivative, and change-based aircraft design. The hierarchy of the aircraft architecture is embedded in present-day processes. Aircraft life-cycle functions follow the classical life-cycle functions. The aircraft-level functions can be flowed to aircraft and subsystem level functions and requirements. Requirements which receive more attention than others include performance, safety, cost, reliability, and weight, not necessarily in that order. Economic requirements include both market-driven and particular customer requirements. Recently published certification guidelines for the aircraft, including its software, incorporate SE principles. Strong SE management is required for the successful development of commercial aircraft.

Systems Engineering for Commercial Aircraft (1997) elaborates on the principles discussed in this chapter.

Commercial aircraft

Although the term 'commercial aircraft' generally refers to jet-powered aircraft carrying large numbers of passengers for long distances, SE principles also apply to freight-carrying aircraft, 'commuter,' aircraft, and 'general aviation.'

There are three types of aircraft developed in the commercial aircraft industry: new aircraft, derivative aircraft and change-based aircraft. All aircraft should meet the top-level requirements, and a thorough verification process is required to show that these requirements have been met.

A new aircraft is developed from a 'blank slate,' using the SE process. The requirements for new aircraft are market driven and reflect the desires of a broad spectrum of potential customers. The top-level synthesis of an aircraft is discussed later in this chapter.

Derivative aircraft are based on previous designs. Like new aircraft requirements, derivative aircraft requirements are based on market-driven needs. The advantage of derivative designs is that significant development costs can be avoided.

Change-based aircraft are those new or derivative aircraft which have been ordered by specific customers and require specific options and custom designs. Often these options can affect aircraft performance and therefore must be evaluated at the aircraft level.

Architecture of the aircraft system

Figure 1.1 shows a typical hierarchy for a commercial aircraft system, which includes more than the aircraft itself. The aircraft is just one of five second-level elements. Secondly, a hierarchical numbering system known as the ATA index and published in the ATA Specification 100 (1989) already exists in the aircraft industry. The correlation between the ATA index and the

A systems approach to developing the new aircraft

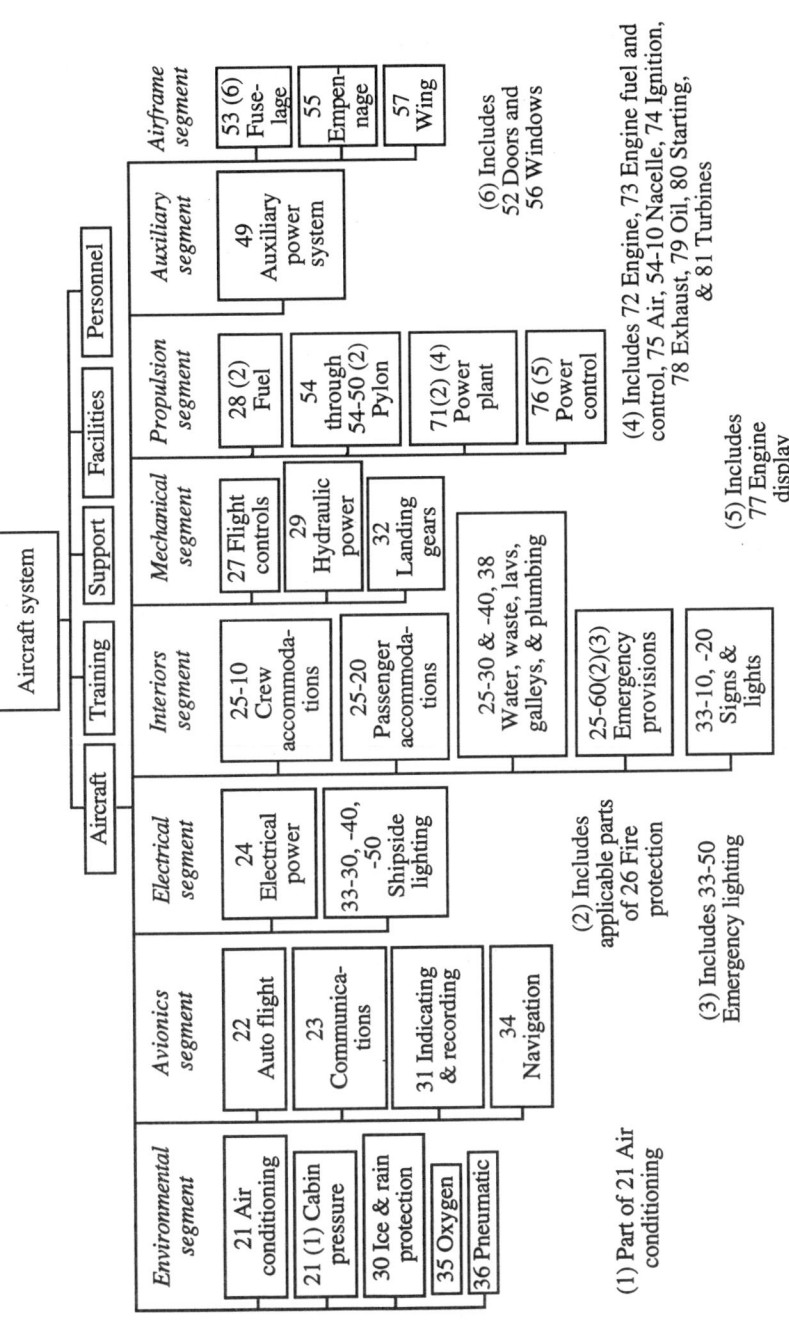

Figure 1.1 Aircraft system architecture (and ATA chapter correlation)

Table 1.1
Comparison of traditional and aircraft life-cycle functions

Traditional SE life-cycle functions	Aircraft life-cycle functions
Development	Market analysis Perform initial marketing Perform initial design Market aircraft Perform design and development
Manufacturing	Perform manufacturing, procurement, and assembly
Verification	Perform design and development Perform certification
Deployment	Operate aircraft
Operations	Operate aircraft
Support	Perform sustainment
Training	Perform sustainment
Disposal	Remove aircraft from service

sample hierarchy is shown in Figure 1.1. This hierarchy is one of many possible valid hierarchies. The development of this hierarchy is one of the first steps in the aircraft SE synthesis process.

Functional analysis

A fundamental premise of SE is that a system and all of the elements shown in Figure 1.1 must meet performance requirements which emanate from a set of functions. By beginning with functions rather than preconceived solutions, the engineer can increase the number of possible solutions. The following sections discuss both the life-cycle and operational functions.

Life-cycle functions

The demands of the aircraft industry give the life-cycle flow its own unique characteristics. Table 1.1 compares aircraft life-cycle functions with the traditional SE life-cycle functions from the draft EIA 632 (1994).

Aircraft functions

One way to identify aircraft functions is to view them as nodes of a matrix. These nodes include the operational phases, mission functions, and situational functions, as shown in Figure 1.2.

Figure 1.3 shows how the Perform Flight Operations functions can be expanded. These functions and their subordinate functions will drive the subsystem performance requirements.

A systems approach to developing the new aircraft

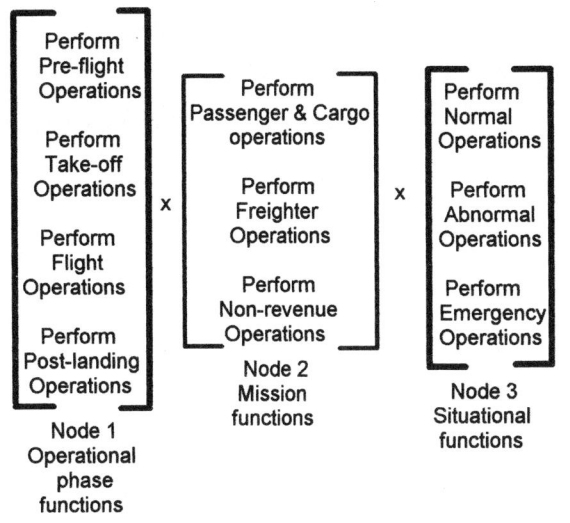

Figure 1.2 Matrix of aircraft operational functions

Requirements

Although sources vary regarding the categories of requirements, the simplest system utilises only two categories: performance requirements and constraints. Each performance requirement is associated with a function and describes how well a function should be performed. For example, at the top level, 'the aircraft shall fly 8,000 nmi.' At the subsystem level, 'the environmental control system shall provide 20 cfm of air through each outlet.' Constraints specify the limits on the performance. Common constraints are dimensional and physical environment constraints. The following subsections discuss the aircraft requirements from the point of view of their sources, namely, economic and regulatory.

Economic requirements

Economic requirements are either market-driven or specific customer requirements. Top-level requirements, such as the range, number of passengers, cargo capacity, and operating costs, will be driven by overall market factors. Specific customers will require special capabilities, for example, relating to interior or avionics design. Requirements can also be initiated internally, such as for a design for manufacturing and assembly (DFMA) initiative.

Regulatory requirements

Regulatory requirements are those imposed by regulatory agencies, such as the Federal Aviation Administration (FAA) and the Joint Aviation Agencies (JAA). The FAA and JAA set the requirements for the certification of the aircraft (discussed later). Many requirements are also set by industry standards and other agencies.

Developing the Future Aviation System

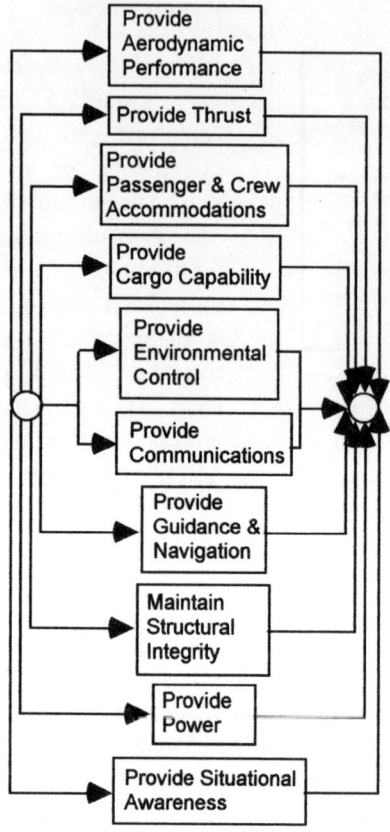

Figure 1.3 The perform flight operations function

Derived and allocated requirements

As the aircraft is developed from the top-level performance requirements and constraints, derived requirements at the top and subsystems levels will be developed. The number of engines and weight of the aircraft are examples. In addition, software requirements, all of which are derived, will be developed.

Many requirements can be allocated to aircraft subsystems. These include, but are not limited to, weight, non-recurring (development) cost, recurring (unit) cost, direct operating cost (DOC), dispatch reliability, maximum allowable probability (MAP) of failure, internal noise (sound levels), external noise, electrical loads, air distribution, fuel consumption, emissions, mainten-ance cost, and loads, shock, and vibration.

Constraints and specialty requirements

The term *constraints* in this chapter refers to all non-performance requirements, that is, requirements which cannot be determined from functions. Whereas performance requirements establish how well a system or subsystem should perform, constraints define the limits on that performance. As we have noted, definitions differ among systems engineers. For example, MIL-STD-961D (1995) uses the term *constraints* for a limited set of design constraints. The present definition, though, is preferred for clarity and simplicity.

Specialty requirements are those requirements set by the various engineering specialties. These include, but are not necessarily limited to, human factors, reliability, maintainability, safety, environments, mass properties, and software.

Constraints and specialty requirements are discussed together here because many constraints arise from the engineering specialties. However, specialty requirements may be either performance requirements or constraints.

Weight

Weight is one of the most closely watched aircraft requirements. The primary weight for design and allocation to subsystems is the manufacturer's empty weight (MEW). The primary weight for sizing is the maximum take-off weight (MTOW). Weight can either be established as a derived sizing parameter or as a constraint from airport weight limits. Either way, weight will be allocated to the subsystems using the classical SE allocation process.

Reliability

Several types of reliability can be applied to aircraft design, but the one with the most visibility is dispatch reliability. Dispatch reliability is the probability that the aircraft will leave the gate within 15 minutes of the scheduled time. Like weight, dispatch reliability can be allocated to all the aircraft subsystems. Safety related reliability is treated within the certification process, discussed later.

Human factors

Although human factors requirements apply to many aspects of aircraft design, such as passenger comfort and maintenance, the area receiving the most attention is flight deck (cockpit) design. The challenge of human factors is the development and allocation of verifiable requirements both to humans and to the equipment, as described by Chapanis (1996). That is, the human is considered part of the system in accordance with the aircraft system hierarchy of Figure 1.1.

The human factors requirements for cockpit design require the resolution of conflicting requirements: On one hand the cockpit must be designed to avoid excessive pilot workload during periods of high stress, such as during landing and during emergency situations. On the other hand, the cockpit must be designed to maintain pilot vigilance during periods of low activity. Another goal of human factors is to minimise the effects of *periferalisation*, that is, the complex psychological state which results from a shift in the pilot role from direct contact and control of the aircraft to one of system monitor, as described by Satchell (1993).

Billings has developed a comprehensive set of requirements for automated flight deck systems. The common thread among these requirements is that

Developing the Future Aviation System

they are all *human-centered*: that is, they focus on what the system must do to facilitate flight crew action and to minimise flight crew error. One example of such a requirement is that 'automation must ensure that operators are not removed from the command role.' The role of the systems engineer, first, is to help the human factors specialist validate that each re-quirement, like this one, is valid for the application at hand. Secondly, after a solution has been proposed for this requirement, an analysis or test must be conceived and conducted to verify that the requirement has been met.

Peg Shaffer, in a later chapter, provides a more complete discussion of the human factors issues associated with aircraft.

Synthesis

Synthesis is the process of converting functions and requirements into actual designs. The selection of the architecture and determination of the functions are the initial steps in the synthesis process. By focusing on the functions and requirements, rather than preconceived solutions, the engineer will leave the door open to new solutions, such as the three-surface configuration or the blended wing-body concept. Hence, the SE process facilitates rather than hinders creativity.

A key part of the synthesis process is the trade-off analysis. For each design choice the engineer must make, for example advanced vs. conventional materials, the solutions must be weighed against the system requirements. For aircraft, the cost and weight requirements are paramount in all trade-off analyses. Although the concept of trade-offs is not new, the focus on system requirements gives new rigor to the design process.

Top-level synthesis

An oceanic system

Feerrar and Sinha (1996) show how the FAA is using the SE approach to develop an Oceanic Air Traffic Management (OATM) system. While the OATM system does not include the aircraft itself, it is an example of a system at the very highest level. In addition, it is a system with which the aircraft of the future must be completely compatible. The system consists of air and ground communications, automation aids, interfacility communications and coordination, and the operating procedures and aircraft separation standards. The system will provide improved communications, navigation capability, decision support systems, and a reduction in aircraft separation.

A systems approach to developing the new aircraft

The FAA has two key techniques in the development of this system: the first technique is the use of managing to the mission. That is, as each element of the system is selected, the system is validated by comparing its performance against a set of scenarios in system simulations. Secondly, the system is being synthesized in a series of three successive builds. In the first build, improved data links are added. In the second build, automated interfaces with other civil aviation authorities (CAAs) are added. In the third build, full automated dependent surveillance (ADS) capability is added for improved communications and decision support. Each build is supported by the synthesis steps trade-offs, risk analysis, rapid prototyping, and evaluation to assure that the mission objectives are met.

Hence, by focusing on the mission objectives and adopting the SE principle that the system must be considered as a whole rather than as a collection of parts, the FAA is progressing towards the goal of developing a new air traffic management system that will meet the needs of the future.

The aircraft system

The aircraft system, as defined in this chapter, consists of the five elements previously discussed. There are many requirements which may affect or be affected by one or more of these elements. Table 1.2 presents a partial list of these requirements.

Aircraft top-level synthesis

The creation and building of complex systems is often called *systems architecting*, as described by Rechtin (1991). Systems architecting goes beyond technical requirements to focus on such concepts as customer satisfaction. We will frame the aircraft synthesis process in terms of performance requirements, constraints, and requirements allocation. The top-level aircraft synthesis process began with the top-level functions (Figure 1.2) and the aircraft system architecture (Figure 1.1).

Aircraft sizing As described by Corning (1977), aircraft sizing begins with wing sizing and balances three conditions: take-off, cruise, and landing. For the purpose of initial sizing, the performance requirements and constraints of Table 1.3 apply.

This process continues with many trade-offs and iterations involving, for example, wing sweepback angle, thickness ratio, etc. The engines are normally sized by the conditions at take-off. However, iterations may size the engines at other conditions. Take-off weight can be estimated from standard sizing relationships involving structure, engines, fuel, payload, and fixed equipment (electrical, hydraulic, environmental control, avionics, etc.). All of

Table 1.2
Requirements affecting the aircraft system

Cargo characteristics	Costs
Airport characteristics	Exterior noise
Utilisation rate	Operational requirements
Turnaround time	Growth capability
People-related requirements	Aircraft autonomy
Passenger service requirements	Consumables
Regulatory environmental requirements	Reliabilities, both dispatch and operational
Configuration change-over times	Particular customer requirements
Actual origins and destinations	

the above information gives us enough information to determine the climb and cruise ranges.

Economic constraints From the descriptions given by Van Bodegraven (1990) and by the author (1995), direct operating cost (DOC) is a primary design constraint. The various components of DOC are navigation fees, insurance, landing fees, ground handling, crew (cabin, cockpit), ownership (depreciation and interest), maintenance (for instance engine, airframe), and fuel and oil.

Figure 1.4 shows how DOC is allocated to the various aspects of the design. In most cases the DOC allocation requires a change in parameter to perform the requirements allocations.

Figure 1.5 shows how DOC is used to select a design point for a new aircraft. Two types of DOC are important: DOC per seat mile and DOC per trip. Design points in the lower left-hand corner are deemed to be economically viable while those in the upper right-hand corner are not.

Subsystem synthesis

Virtually all subsystem requirements are derived requirements. The flow-down of requirements from the top level is dependent on the architecture selected at that level, such as the one shown in Figure 1.1. The subsystem performance requirements are also driven by the subsystem functions identified in Figure 1.3. While each of the individual segments shown in Figure 1.1 may derive its requirements from individual functions, many functions apply to several segments. Hence, SE trade studies should be conducted both between and within aircraft segments and subsystems. For example, the weight savings of advanced material would be traded against the increased costs. In addition, the trade-off is key to the introduction of advanced technologies discussed below.

New technologies and concepts

To meet design requirements for reduced weight, noise, and emissions, robust systems, and safe and economic operation, many advanced technologies are routinely incorporated into commercial aircraft, e.g., heads-up displays (HUD), voice recognition, global positioning system (GPS) receivers, point-to-point inertial navigators, reconfigurable instrument displays based entirely on digital video displays, Doppler radar, fly-by-wire (FBW) or fly-by-light (FBL), and real-time computer fault detection and isolation. Composite material technology is key to weight reduction.

Radical changes in aircraft design are being studied. For example, the drive towards a high speed civil transport (HSCT) has focused on advances in propulsion and materials. The three-surface aircraft described by Martínez-Val (1994) has been shown to improve the aerodynamic performance. In addition another intersting concept to improve aerodynamic performance is the blended wing-body (BWB). This concept resembles a large manta ray.

Table 1.3
Wing sizing requirements

Performance Requirements	Constraints
Number of passengers	Field length
Weight of cargo	Initial cruise altitude
Range	Atmospheric conditions
Cruise Mach number	Approach speed

SE has the capability of evaluating the introduction of advanced technology for both subsonic and supersonic aircraft, as described by Mackey (1996). Each technology must be evaluated as part of the trade-off and risk analyses which are part of the SE synthesis process.

Interfaces

Interfaces are the connections which make the many elements of a system interact with each other, whether they be elements of the global aviation system or the aircraft itself. SE emphasizes both the functional and physical interfaces. The functional interface describes the purpose of the interface, while the physical interface characterizes the dimensions of the connection.

Developing the Future Aviation System

Because aircraft components are developed in various parts of the world and brought to a single location for assembly, adherence to the SE interface principles is even more important. Aircraft functional interfaces include electrical power, hydraulic power, pneumatic power, mechanical forces and torques, conditioned air, heat, vibration, shock, loads, and signal interfaces. The SAE Generic Open Architecture Framework establishes nine classes of interfaces. This framework promises to standardise interfaces and make interface development easier.

Traditional aircraft development fixtures (DFs), that is, aircraft mock-ups

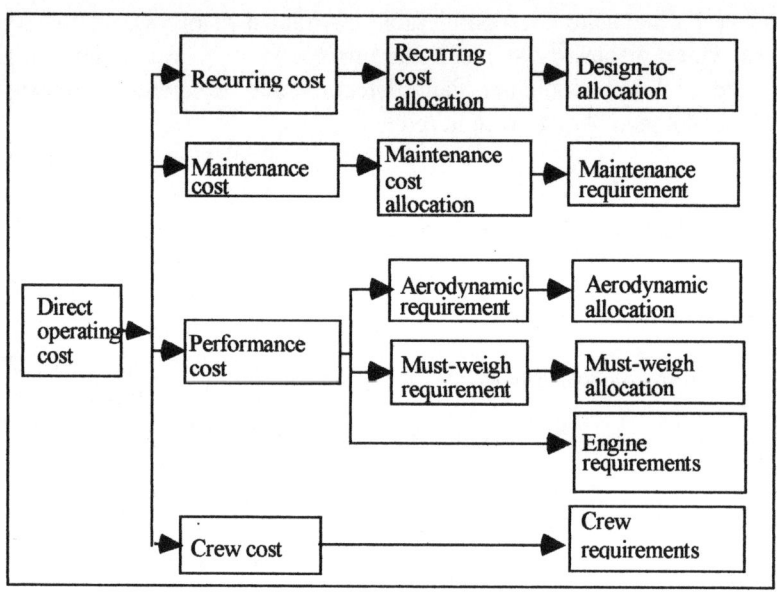

Figure 1.4 Allocation of direct operating costs

to coordinate physical interfaces, are being replaced by electronic development fixtures (EDFs), that is, three dimensional electronic models. This trend promises to reduce the cost of aircraft development, speed the development cycle, and improve the quality of interfaces.

As the global aviation system evolves, interfaces between the aircraft and the airport facilities and communications systems will increasingly be affected. The blended wing-body concept, for example, will require significant changes in passenger access, fuelling techniques, and aircraft stationing. Future Air Navigation (FAN) systems will require new communications interfaces. In the SE process these interfaces will be considered before, rather than after, the aircraft are designed.

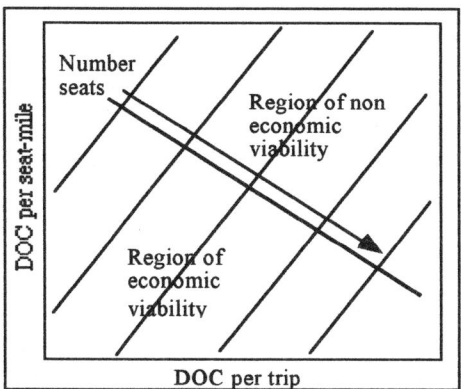

Figure 1.5 DOC regimes

SE in the manufacturing process

SE is often considered to be an up-front process, that is, a process which ends when the design begins. Nothing could be further from the truth. Apart from the verification activities which occur during the whole development cycle, SE can have a major impact on the manufacturing process.

From the beginning, SE has required developers to identify any specific manufacturing requirements which may be required for any system element. For example, these requirements may include manufacturing tolerances or the use of special materials.

Other less recognized manufacturing involvement includes manufacturing as a source of requirements. In addition to the traditional requirements sources (customer, regulatory agencies) the requirements process should allow for the manufacturing community to provide requirements which will improve product quality or reduce cost.

Another misperception is that SE only develops requirements for the product, that is, the aircraft and its components. In reality, SE also develops requirements for the manufacturing process. An example is the use of dimensional management (DM). In DM, trade-offs are conducted which determine the combination of assembly and tooling tolerances which will result in the product which meets its design tolerance requirements.

Certification

Certification is the process that substantiates that the aircraft and its subsystems comply with airworthiness requirements. The FAA and the SAE have taken a major step towards incorporating the SE process into the certification process with the publication of ARP 4754 (1996).

Certification encompasses the entire development of the aircraft. Required documentation includes descriptions of aircraft development, requirements and their validation and verification, system integration, configuration management, and process assurance. All of these activities are completely in line with the SE process.

Safety analysis

Safety is the primary focus of certification. Specific analyses to assure the safety of the aircraft are treated in accordance with SE principles as shown in Table 1.6. The principal safety analyses are the functional hazard analysis (FHA), the preliminary system safety analysis (PSSA), the system safety analysis (SSA), and the common cause analysis (CCA). This figure shows how the safety assessment and SE processes are linked.

Table 1.6
Compatibility of the SE and certification processes

SE elements	Certification aspects
Functional analysis	Aircraft-level functional requirements, allocation of aircraft functions to systems, functional hazard analysis (FHA).
Requirements development and allocation	Requirements categories, allocation of item requirements to hardware and software, preliminary safety assessment (PSSA), validation plan and data
Synthesis	Development of system architecture, common cause analysis (CCA), system implementation
Verification	Verification data, system safety analysis (SSA), inspection and review process
SE management	Certification plan, development plan, configuration management plan and data, process assurance plan and evidence, certification summary

Software development

Another area requiring adherence to certification requirements is software development. The process, described by RTCA/DO-178B (1992), for the development of the software itself is essentially the same as the SE process. Secondly, the certification process does not consider the software to be a separate entity, but rather a part of a larger system to be certified. Although software is only a part of the total system, SE owes much of its development to principles developed within the software community.

SE management

The important aspects of SE management for commercial aircraft include, first, the conduct of rigorous design reviews, particularly at the aircraft level. Secondly, the use of integrated product teams (IPTs) is essential to develop, implement, and verify requirements for each of the major aircraft segments shown in Figure 1.1. A supplier management process which includes the suppliers as part of the IPTs is important. IPTs also assure that normally down-stream processes, such as maintainability, get included at the beginning. Thorough configuration management is essential for aircraft integration. Risk management is essential especially for the introduction of new technologies. Finally, it is important for SE management to assure that safety is not compromised by organisational factors, as described by Paté-Cornell (1990).

Conclusions

Many changes in aircraft and subsystem concepts are being studied by the aircraft industry. Cockpits are being studied with the goal of improvements in safety. All of these developments will be enhanced by the application of SE during the development cycle.

References

Air Transport Association (ATA) (1989), *Specification for Manufacturers' Technical Data*, Specification 100, Revision 28.
Billings, Charles E. (1997), *Aviation Automation: The Search for a Human-Centered Approach*, Laurence Erlbaum Associates, Mahweh, New Jersey, p. 246.
Chapanis, Alphonse (1996), *Human Factors in Systems Engineering*, John Wiley and Sons, Inc., New York, pp. 206-07.

Corning, Gerald, *Supersonic and Subsonic* (1977), *CTOL and VTOL, Airplane Design*, published by author: College Park, Maryland, pp. 2:1-102.

Department of Defense (1995), *Defense Specifications*, MIL-STD-961D. (supersedes MIL-STD-490A)

Electronics Industries Association (EIA) (1994), *Systems Engineering*, EIA IS (Interim Standard) 632. (draft)

Feerrar, Wallace N. and Sinha, Agam N. (1996), 'Collaborative System Engineering Approach to an Evolutionary Oceanic System Development,' *INCOSE Proceedings*, pp. 45-50.

Jackson, Scott (1995), 'Systems Engineering and the Bottom Line,' *Proceedings of NCOSE*, pp. 545-49.

Jackson, Scott (1997), *Systems Engineering for Commercial Aircraft*, Ashgate Publishing Limited: Aldershot.

Mackey, Dr. William F. (1996), 'Conducting a Technology Management Assessment,' *INCOSE Proceedings*, pp. 153-62.

Martínez-Val, Rodrigo, *et al.*, (1994), 'Design Constraints in the Payload-Range Diagram of Ultrahigh Capability Transport Airplanes', *Journal of Aircraft* 31(6), pp. 1268-72.

Paté-Cornell, M. Elisabeth (1990), 'Organizational Aspects of Engineering System Safety: The Case of Offshore Platforms,' *Science* 250, pp. 1210-16.

Petersen, T. J. and Sutcliffe, P. L. (1992), 'Systems Engineering as Applied to the Boeing 777,' AIAA 1992 Aerospace Design Conference, Irvine, California.

Rechtin, Eberhardt (1991), *Systems Architecting: Creating and Building Complex Systems*, Prentice-Hall: Englewood Cliffs, New Jersey.

RTCA, Inc., (1992), *Software Considerations in Airborne Systems and Equipment Certification*, RTCA/DO-178B.

Satchell, Paul (1993), *Cockpit Monitoring and Alerting Systems*, Aldershot: Ashgate Publishing Limited, p. 10.

Society of Automotive Engineers (SAE) (1996), *Guidelines for the Certification of Highly-Integrated and Complex Aircraft Systems*, ARP 4754.

Van Bodegraven, George W. (1990), 'Commercial Aircraft DOC Methods,' AIAA publication AIAA-90-3224-CP.

The author

Scott Jackson, MS, MA studied aeronautical engineering at the University of Texas at Austin. As a summer intern, he worked on the Viscount aircraft at Vickers Aircraft Ltd. (now part of British Aerospace) in England. He later earned a master's degree at the University of California in Los Angeles, with an emphasis on fluid mechanics. His thesis, 'On the Theory of Magnetohydro-dynamic Shock Waves' (1966) was an examination of MHD shock waves in fluids of varying conductivity.

Mr. Jackson is a Senior Principal Specialist in Systems Engineering at McDonnell Douglas in Long Beach, California. He has over thirty years experience in systems engineering in both space and aircraft development. He joined the Douglas Aircraft Company in 1957 as an aerodynamicist. His AIAA technical note, 'Special Solutions to the Equations of Motion for Maneuvering Entry' (1962) was a seminal solution for Shuttle Orbiter type reentry trajectories.

Following the merger of the Douglas and McDonnell companies in 1967, Mr. Jackson devoted his energies to the study of the systems engineering approach. At the McDonnell Douglas Astronautics division, he was Chief Systems Engineer on a DoD project.

As a member and officer of the International Council on Systems Engineering (INCOSE), Mr. Jackson has authored four papers: In 'First-Order Systems Engineering: A Case Study', (1993) he shows how the systems engineering approach can be applied to the development of a lunar base. 'The Morphological Approach: Its Role in Systems Engineering and Its Application to Solar Energy Conversion' (1994) demonstrates a technique for examining many approaches to solar energy conversion. 'Systems Engineering and the Bottom Line' (1995) shows how the systems engineering process can enhance the financial performance of an organisation, using examples from the commercial aircraft industry. 'Introducing Systems Engineering Into A Traditionally Commercial Organization' (1996) provides a set of guidelines for systems engineers to use in the commercial world.

His book *Systems Engineering for Commercial Aircraft* (1997) was published by Ashgate Publishing Limited. A paper by the same name was presented to INCOSE in 1997.

In addition to his technical work, Mr. Jackson has found time to collect two degrees in Spanish. His thesis 'The Union of Mathematics and Poetry in the *Purgatorio* of Raúl Zurita' (1987) was published as the introduction to the English translation of the Chilean poet's work. Another paper 'Juan Luis Martínez: The Novelist of Non-Existance' in the journal *Confluencia* (1996) shows how another Chilean poet uses mathematics as a metaphor.

2 New generation airports

Jenny Beechener

The challenges facing airports in the 21st century

The millennium marks a watershed in the way airports do business and market themselves in an increasingly tough economic climate.

You can count on the fingers of one hand the number of new airports that will open up in the next decade. Greenfield sites in the Asia Pacific region are among the largest, where offshore projects on man-made islands offer relief from some of the constraints imposed on existing airports. The growing environmental lobby and restrictions in land use demand more innovative solutions to the capacity issue worldwide.

The US$25 billion Chek Lap Kok development opens the world's largest passenger terminal in Hong Kong in 1998, with capacity to handle 80-90 million passengers a year by 2010. A new airport at Kuala Lumpur, a second Bangkok airport and the offshore project to build the New Seoul International airport will help meet the region's growing traffic.

China has embarked upon an ambitious plan to upgrade 40 airports to cope with predicted rise in passenger numbers from 60 million 1995 to over 250 million by 2010 (see Table 2.1).

These new sites will go some way towards meeting predicted demand growth of some six per cent annually (eight per cent in south east Asia). However they fall woefully short of establishing a transport network that will cope with 21st Century traffic needs. Nowhere is this more apparent than in Europe.

Only two new airports will open in Europe in the next decade: Oslo and Athens. Others such as Milan-Malpensa, Berlin and Madrid are adding new terminal and runway complexes almost as large as new airports. New runways are planned at Paris-CDG and Amsterdam Schiphol. In the United States, no new airports are planned in the foreseeable future, and new projects face tough funding and regulatory hurdles.

What is becoming clear to the millennium airport is the need to reconcile diminishing resources with growing demand. Those airports that understand these forces will be the future survivors, and the successful hubs of the future.

Privatisation

1997 was something of a watershed year for airport management as Table 2.2 shows. The trend to private ownership accelerated with the acclaimed US$2.6 billion sale of three major Australian airports. South Africa, Argentina, Mexico, Brazil, and Germany have since invited private investors to take an equity stake in the airport business. "Airports are being nudged, or more properly stated forced, into redefining themselves for survival's sake," explains Manfred Scholch, deputy chairman of Frankfurt airport. The year saw explosive growth in airport privatisation initiatives and a rush to borrow money from commercial markets.

Privatisation will not just open up new revenue streams, it will allow financially strong airports to buy smaller airports, creating symbiotic networks. Scholch says the millennium airport "will no longer be a single-site restricted-activity, passive management operation...but an exporter of its staff expertise and professional experience in the development and profitable operation of other world airports." In this scenario, a core of five or six "pace-setting" airports will lead the global market at the start of the new century.

The Frankfurt approach to the future is to accept that no major hubs will be built. Instead, more efficient and commercially responsible policies will provide enough capacity for the coming years from current facilities. The traditional concept of the state-run airport is giving way to a global industry, driven by market forces and liberalised policy making.

Liberalisation in Europe

In much the same way liberalisation has swept through the airline industry, deregulation has introduced more competitive practices by airport operators. Revenue streams have widened to embrace retail and property services, in some instances contributing over 60 per cent to total airport income. In January 1998 the European Commission's directive on ground handling services took effect.

Deregulation of ramp services is welcomed by the airlines as a much overdue measure to bring fairer pricing to the airport. Blamed for driving up prices, and in some cases depressing service standards, ground handling monopolies have been criticised by the airlines for years - claiming charges at some European airports are 30-40 per cent higher than they should be. The situation arises where the airport authority or national airline is the only service provider, with the worst examples in Germany and Mediterranean countries.

Cranfield University Air Transport Group reports widely differing ground handling costs between airports. Charges at Frankfurt, where the airport is the only company responsible for ground handling, are twice those at most other airports and three times those at Manchester and London Gatwick where several companies provide handling services. Turnround costs for a Boeing 757 at

Frankfurt exceed US$3,000 yet fall well below US$2,000 at Manchester and even below US$500 at Larnaca, Cyprus. 'There is clear pattern that monopoly handling airports are generally more expensive', says the report, 'with a correlation between the handling charge and the number of service providers'.

The directive allows airlines to do their own in-terminal passenger handling. Where airports have more than one million passengers or 25,000 tonnes of freight, airlines have the right to self-handle baggage and ramp operations. From 1999, at airports with more than three million passengers or 75,000 tonnes of freight, airlines and handling companies gain the right to offer third party services. From January 2001, at least one of these organisations must be free from direct or indirect control by the airport operator or by a dominant airline (one that carries more than 25 per cent of the airport's passenger or freight traffic).

While the directive is intended to liberalise access to the ground handling market, its impact is curtailed by a host of exclusions and limitations. German airports are among the first to exploit the generous transition period written into the directive, while small airports remain outside the legislation. The earliest that a third party handler can break into a monopoly market is 1999, and member states still have the power to limit the carriers permitted to provide services.

The picture is complicated by further conditions. Member states may justify monopoly management on the grounds of complexity, cost or environmental impact. Airport rules may be imposed on independent service providers as to law and order, safety and security. Protection of workers' rights and "respect for the environment" leads to uncertainty and gives member states and airport authorities a wide latitude to introduce as little liberalisation as possible.

While slow in arriving, the competition rules apply to all member states and affected parties can bring actions in national courts alleging infringement of the rules. The signs are that private initiatives of this kind may prove the faster route to real liberalisation in the ground handling market in Europe.

The handling industry estimates that of about 50 airports that handle more than two million passengers annually, some 30 fail to meet the competitive criteria set down by the directive. Europe's largest independent handling company Servisair identifies new business opportunity, but is hampered by slow implementation. Chief executive John Willis says, "airside monopolies, which are the prime objective of the directive, have exemptions up to four years." In some cases companies are more protected than before, eligible for six year extensions while free to invade open competitive markets elsewhere.

There are early signs of the market opening up. At London Heathrow where ramp and passenger handling has been restricted to third-party services by eight approved airlines, Servisair became the first independent handling company to provide services in 1997. The airport operator, Heathrow Airports, had argued that space constraints in congested terminal areas and safety and security reasons justified its restrictive policy. The Monopolies and Mergers commission saw the concession in the spirit of the handling directive.

The Commission is also tackling the issue of airport charges. According to ICAO figures, airport charges represented 4.3 per cent on average of airline operating costs in 1995. In Europe, they represented 8.5 per cent of costs of members of the Association of European Airlines. Moreover, the charges varied for an Airbus 320 from Ecu 800 at Bilbao, to Ecu 1,700 at Madrid and Ecu 4,800 at Vienna.

The Commission proposes equal treatment for all users, transparency, and a clear relationship between cost and charges levied. Daniel Jacob, deputy head of Commissioner Kinnock's private office explained plans to introduce consultation procedures between airports and users. "We wish to enable airports to contribute to a better management of existing capacity. This means airports should continue to modulate charges in order to deal with congestion. They should also be able to vary charges in order to encourage use of quieter and less polluting aircraft and in order to deal with night time flights."

The measures are expected to reduce airline costs and help airports deal with congestion issues, with adoption by the Transport Ministers targeted in the second half of 1998.

A final part of the Commission's policy to liberalise the air transport sector relates to slot allocation. Present rules give priority to new entrants while preserving the principle of grandfather rights. Market forces are expected to play a larger role in any revised legislation on slots. The Commission proposes independent coordinators, and suggests up to 50 per cent of slots be allocated to new entrants.

Daniel Jacob points out that the creation of a slot market could be done in a way to ensure transparency of transactions, prevent airlines blocking competitors, and guarantee competition by capping the percentage of slots that a particular airline could buy at a given airport. While a proposal is imminent, the issue gathers urgency. As Jacob says, "In 1995 some 700 million passengers used European airports and over nine million movements took place. On the basis of annual growth rate of five to six per cent a year, it is likely that the air transport system will have to cope with a doubling of activity over the next 10-15 years. If the airports face serious challenges, they are also offered formidable opportunities."

Challenges in Asia/Pacific

Nowhere are the challenges more apparent than in the Asia Pacific. With the highest growth rates in the world, and forecast to handle over 1.1 billion passengers in 2010 (half the world's total), the region is busy investing in new facilities. The World Bank estimates aviation infrastructure expenditure between 1995 and 2010 will reach US$200 billion by Asia-Pacific states. As governments are increasingly unwilling or unable to finance this development, private capital is likely to bear the brunt of this expenditure.

Developing the Future Aviation System

The region has begun construction of series of mega-airports to meet the level of passenger traffic activity. In 1997 the top six airports worldwide by passenger volume were Chicago O'Hare (69 million), Altanta Hartsfield (63 million), Los Angeles International (58 million), London Heathrow (56 million), Dallas-Fort Worth (55 million) and Tokyo Haneda (47 million). Asia-Pacific airports do not compare in passenger volume with the exception of Tokyo Haneda. The top six airports in 1997 after Haneda were Seoul (35 million), Hong Kong (30 million), Tokyo Narita (25 million), Bangkok (25 million) and Singapore (25 million).

Gate-to-gate concepts

The world ranking will change dramatically in the next decade. The new Asia Pacific airports will have the highest passenger handling capacities in the world, while European and North American facilities are approaching capacity limits of existing infrastructure. At this point, Europe may reveal some useful lessons in how to expand capacity without drawing heavily on new resources.

One area where striking capacity growth is already evident is airspace and air traffic control. The European programme set up to harmonise the region's airspace expects to expand capacity by 70 per cent by 2003. Alex Hendriks, head of airspace and navigation at the Eurocontrol organisation predicts, "the overwhelming cause of delays will not be air traffic control inefficiencies, but a lack of airport runways and terminal gates."

Results collated by the central office of delay analysis contrast three airports: Athens, Frankfurt and London Gatwick. Only 5-10 per cent of delays at these airports are due to en route air traffic control. The remainder is shown to be due to a combination of weather, baggage and other ground services. Hendriks warns solving these issues are not straightforward: "How profit-making airports can coordinate with non-profit-making air traffic management organisations to produce a gate-to-gate concept will be very difficult."

Important capacity-enhancing measures are available, ranging from reducing vehicles on the ramp, to improving runway utilisation, adding high-speed exits, new queuing strategies and approach/departure enhancement rules. They are not sufficient to produce the huge savings in time and space that will be needed over the next decade. For that, it will be necessary to introduce new concepts in terminal passenger handling and design.

The goal is to produce a seamless travel experience, within which passengers can move effortlessly through the terminal from kerbside to aircraft gate without the bottlenecks of check-in, border control, security checks, duty-free and departure lounge boarding. Automated ticket and boarding pass technology promises the platform for a quantum leap in passenger processing.

New generation airports

The key will be to put the passenger at the centre of a new generation of electronic travel processing system. From reserving a ticket via the Internet (saving around US$8 a time) to boarding the aircraft, a single integrated electronic network will link self-service ticket kiosks, airport shopping centres and cash machines.

Instead of the passenger's travel data being stored on separate airport and border control databases, the passenger will carry all this information on a single smart card which is electronically readable at the airport area gateways. What is more, all this can evolve as a result of adaptations to existing technology.

Though few would admit it, the problem facing all the world's congested airports is that increased commercialisation and exotic new information technology are exciting solutions to the short term problems of a growing market, and industry leaders in other sectors would give their eye-teeth to face the problems of relentless demand for their services.

Table 2.1
Major new airport developments in China (PRC)

City/Airport	Province	Value (US$m)	Completion
Guangzhou/Huadu	Guangdong	2,400	2000
Shanghai/Pudong	Shanghai	2,350	2000
Hangzhou/Xiaoshan	Zhejiang	370	
Haikou/Meilan	Hainan	340	1998
Fuzhou	Fujian	315	
Guilin/Liangjiang	Guangxi	175	1996
Nanjing/Lukou	Jiangsu	130	1997
Nanchang	Xinjiang	128	
Xuzhou	Jiangsu	106	1996
Yinchuang/Hedong	Ninxia	65	1997
Yichang/Sanxia	Hubei	46	1996

Table 2.2
Timetables for privatisation

1987
BAA privatised

1992
Vienna partial privatisation
Athens planning privatisation

1994
Privatisation planned or underway at:
Belfast
Istanbul
Copenhagen
Australia and Germany set out privatisation agendas.

1996
Privatisation planned or underway at:
Rome
Australia outlines its stage two privatisation plans and India draws up a privatisation blueprint.

1997
Privatisation planned or underway at:
Argentina
Australia (stage two)
Brazil
Germany - Berlin, Dusseldorf
UK - Bristol, Birmingham, Luton
South Africa

Reference sources

Jane's Airport Review
Jane's 1997 conference: The Millennium Airport
ICBI's Global Airport Development Financing & Marketing 1997 conference
Jane's Special Report: Global Airport Expansion

The author

Jenny Beechener is the Editor of Jane's Airport Review.

3 The airport business and information technology

Mark Sawle-Thomas

Airports have a very high profile. The major international airports may be seen by many as necessary evils and polluters of the environment, but they are very real generators of work and economic impetus. Smaller airports are generally symbols of civic pride, centres of commerce and vital elements in the transport network enabling rapid mobility for far flung communities, although some may consider them to be unwelcome intrusions into the peace of the area. Love them or hate them, airports are an essential component in an ever expanding air transport industry: there appears to be no let up in the continuing increase in the number of people wishing to travel by air, and the air cargo industry, too, is seeing unprecedented growth.

The pressure on capacity at the major airports, the growing service expectations of airport users, and the reluctance of governments to continue to provide public funds are but three of the pressures facing airport managers today. While Information Technology cannot provide all the answers, it can alleviate some of the problems and provide the foundation for striking growth.

This chapter reviews the role of airports in the modern world, and considers how the advances in information management processes may be deployed to improve the lot of airport managers and staff, and their customers - both other airport user organisations and passengers.

The role of airports in the world

As airports take over from the sea ports as the major confluence of routes for the carriage of people and goods, their place in the conduct of world trade becomes increasingly prominent. An additional very real stimulus for the growth of their importance in national life is the trend towards commercialisation: the search for profit and the introduction of competition. The change from airports as landing fields to airports as big business has been driven by the change of many from state, or local authority, to private

Developing the Future Aviation System

ownership. Even where the regional authority has maintained a controlling or significant interest, many airports are being treated as commercial enterprises with a profit motive.

In 1996 EDS undertook a world-wide survey of leading airports, airlines and academic institutions in order to gain a clearer picture of how the airport industry was developing. Extracts from the White Paper, 'Airports Look to the Future', produced as a result of this survey are included in this chapter. The study looked, *inter alia*, at the place of an airport in its regional economy. The report states:

The trends identified ... cast a broad shadow over all the challenges that airport directors face—including building revenue, expanding service, developing innovative, mutually beneficial alliances, and knowing what it is [they] need to know.

... Those airport executives who can take a broad view and then develop strategies within its context will increase their chances of pulling away from the pack and achieving both short- and long-term advantages. The ripple effect [of the airport on the business life of its community] *can be ... pictured in a circular fashion as shown in Figure 3.1.*

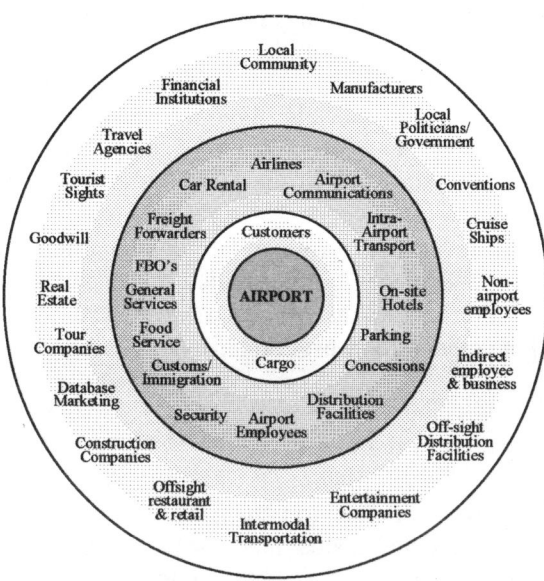

Figure 3.1 Business effects of an airport

Some of the findings were:

Analysis of Air Transport Action Group (ATAG) data suggests that every passenger flowing through a major airport generates as much as US $1,600 in total revenue per trip.
According to additional analysis of ATAG data, one local job is created for every 125 passengers that pass through an airport.
Cargo also contributes to community employment rates: Airports Council International (ACI) estimates that one job is generated for every US $100,000 in cargo revenue; a large airport hub might easily account for $5 billion to 10 billion in cargo shipments.

Commercialisation and privatisation of airports

BAA was among the first to privatise in 1987, and it remains a paradigm for airport management - Aer Rianta and Schiphol are examples of airport companies which, although state owned, operate as commercial enterprises, and both of these are at the forefront of airport business. It is to these and other European airports, where the thrust for change originated, that the growing airport sector in the Asia Pacific region looks for example, advice and in some cases participation. Their influence is being felt now in the USA where until recently airports tended to be little more than landlords for airlines.

The EDS White Paper included the diagram, shown in Figure 3.2, which illustrates the current management position of airports across the world.

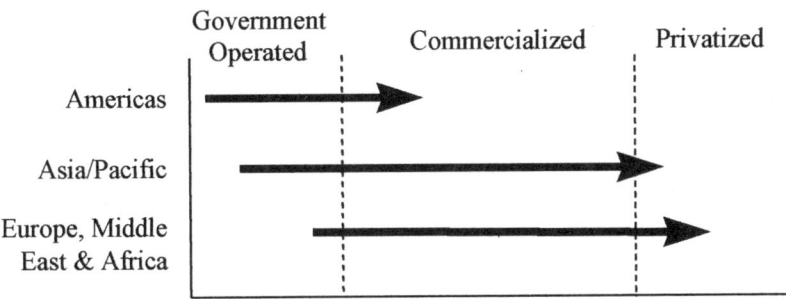

Figure 3.2 Current management position of airports across the world

Key findings of the report were:

An increasing number of major hubs in Europe are privatized.

New airports in Asia tend to be equal partnerships between business and government.

The United States has no fully privatized airports; the major hubs, however, are beginning to commercialize.

In a related control issue, the structure of airline lease agreements is undergoing radical change at airports across the United States.

The days of the traditional, long-term lease agreement appear to be numbered. Compensatory rates are taking the place of residual charges.

Terminals are moving from single-airline ownership to shared facilities/common use.

What has become clear over these last few years is that an airport can have a significant measure of control over its own destiny. Certainly geography and history play a part, but these alone do not explain the success of some airports over others. Manchester Airport in the UK is an example of the close co-operation between an enlightened local authority and a dynamic management team leading to growth well beyond most of its peers.

The more progressive airport companies now realise that they must look outside their own home ground if they are to grow further and provide an attractive investment opportunity. Airport alliances are being forged to provide a consistency of service and a network of interconnecting routes; international groups are being created to enable economies of scale and to broaden the business base; additional non-aviation revenue generating activities are being developed, such as airport associated business parks and entertainment complexes; and there is a continual search for new means to provide improved services for airport users and the surrounding communities.

Airports Look to the Future states:

The most progressive airports have recognized that consumer information has become the currency of experience marketing. This information is the key to understanding buying patterns, which, when placed under the analytical microscope, become understandable and predictable. Airports have a unique opportunity to open a new door on the travel experience, attain a competitive edge in the marketplace, and find new streams of revenue.

The key findings were:

> About 70 % of survey respondents said incremental revenue will be the most important issue over the next five years.
>
> Airport directors, specifically, chose "Increase Revenues" (from a diverse list of business issues) as the most important issue facing them.
>
> Revenue is shifting from aeronautical sources to non-aeronautical, with the degree depending on each individual airport's strategy.
>
> "Traditional" airports' revenues are split about 58 % aeronautical and 42 % non-aeronautical.
>
> "Active" airports' revenue is now split about 50-50 between aeronautical and non-aeronautical.
>
> "Aggressive" airports are realizing a revenue split of about 66 % non-aeronautical and 34 aeronautical.

The pie charts, shown in Figure 3.3, illustrate the proportion of non-aeronautical revenue generated by differing approaches to marketing airport services.

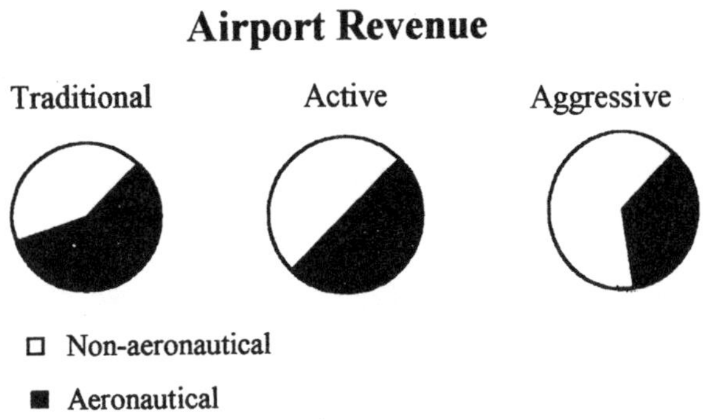

Airport Revenue

Traditional   Active   Aggressive

☐ Non-aeronautical
■ Aeronautical

Figure 3.3 Proportion of non-aeronautical revenue at airports

Multi-modal travel

There is also a growing awareness of the place of air travel in the overall transport spectrum. As the point at which air and surface travel meet, airports are linchpins of the multi-modal travel system. Amsterdam Airport Schiphol, for example, regards itself as a 'mobility manager', aiming to be the focus for travellers and providing door-to-door travel information, including road and rail connections. With the growth of high speed train networks, particularly in Europe, it may be that the journey will not even include an air sector. This creates a fundamental shift in relationships: the airport becomes the primary provider. The traveller is the airport's customer and the airline supplies the airport. Airports will be seeking to provide a balanced routing pattern and convenient transfers to a series of preferred destinations and they will be seeking to allocate their scarce landing slots to those carriers willing to co-operate in the realisation of their strategy. Airlines will not be able to call all the shots in the way that they once were. This shift in emphasis requires a fundamental change in the approach to management, and the services that support it - among them Information Technology.

An airport is many businesses

Being a staging post for air travellers and freight has become only part of the modern airport business, which usually also encompasses at least property management, retail operations, construction and security, as well as the new business ventures outlined earlier. As the primary function of the airport, flight operations have a significant influence over everything else that happens, but this only serves to emphasise the need to consider aeronautical activity as an element of an integrated whole.

While the multiple business strands may operate more or less autonomously, they are generally designed to support each other by drawing new customers to the airport, and by providing additional attractions to enhance the services provided to existing customers. The common strands are the customer base, different customers will make use of different combinations of services, and the airport corporate management which is seeking to increase market share and optimise the profit potential of the airport facilities.

The IT picture should reflect this interdependence, but unfortunately this is too rarely the case. Frequently, information is organised on departmental lines - with separate systems often duplicating functions of those managed in other areas. Information from one system will be re-entered into others, and it is difficult for management quickly and easily to obtain comprehensive reports for analysis and planning. Also, it may be that the organisation has a variety of office management systems making communication between business elements, or even within businesses, more difficult than is wise or necessary.

Systems Integration is only part of the answer

Systems Integration is sometimes put forward as the answer, and, in part, it is. If existing systems are enabled to share data, much of the duplication of effort will be eliminated, but the nature and limitations of the systems may well be a constraint on the structure of the organisation and the management of its processes.

A mistake that is sometimes made when considering systems integration is to view it in isolation. It is merely a tool to facilitate the cohesion of a business and the integration of business processes. It is just one strand of a much more complex process of concentrating an organisation's resources to focus on the long term objectives. Other strands may include: business process re-evaluation, human resource deployment, building layout, the physical environment, and support and maintenance.

IT in support of business development

One leading airport has estimated that it will improve efficiency by over 20% through implementing integrated supporting IT services and re-aligning its business processes.

Increasing technological sophistication means that the application of IT allied to innovative thought can have a fundamental effect on the way a business is run, and can have a tremendously beneficial effect on profitability and/or market share - particularly important for airports as they move forward as business enterprises.

In any review of the role of information management, it is necessary first to establish the strategic direction the airport wishes to take. This will lead naturally to an analysis and review of businesses processes designed to achieve the objectives. New business processes are likely to generate a review of the organisation established to carry them out, with, perhaps, new posts and new responsibilities. A definition of the information needs of the new organisation and its staff will be a natural extension.

EDS is one company that has established an organisation, under the title of 'Enterprise Solutions', specifically to address these types of issues. It combines the consultancy skills of its subsidiary, A. T. Kearney, with the more traditional process management and systems development skills of the parent company. This multi-disciplinary approach has had dramatic effects on productivity and profitability for several customers.

Information management is a business issue

A key point to recognise is that information management and the systems or services that support it are business issues which needs to be viewed at the highest level. They are not departmental issues. Information will not recognise the artificial boundaries all organisations put in its way, nor is it sensible to view costs and benefits on a departmental level since these too may cross boundaries, and in an uneven manner, making their allocation in a traditional fashion difficult if not impossible.

Only by understanding the whole business picture can the information needs of the various departments be assessed and compared. Common threads will become apparent, the potential for shared systems may be exploited, gaps in coverage may be filled and duplication may be eliminated or at least kept to a minimum. More importantly, the provision of information may be geared to meeting strategic objectives that are not apparent at departmental levels.

Through dialogue with suppliers based on driving towards these desired objectives, rather than on delivering specified systems, an airport should be able to identify the most cost effective means of satisfying its business need - which in many cases will achieve far more than expected, but perhaps in a totally unexpected manner requiring organisational change.

Airports should *expect* IT companies to respond at this business level, and they will find that potential suppliers will have much to contribute from previous experience in the travel and other industries. By identifying the nature of the problem and the desired result the airport management may draw on this experience. In order to avoid some of the pitfalls of the past which resulted in misdirected purpose, late delivery and poor performance, it is preferable that the procurement of IT systems should be collaborative rather than confrontational.

With this approach, the boundaries between operational, engineering and business systems become blurred - they feed off each other. Communication between people and systems, rather than the mere abstraction of data, becomes the driving force. Information Management becomes a saleable commodity - IT as a revenue generator rather than a cost: a revolutionary thought! Old hierarchies and divisions break down - co-operation, self motivation and initiative become more important than command and control.

Technology no longer constrains business in the way it once did: it now provides the freedom for managers and planners to take the broad view of an organisation and demand that whatever real information is wanted is made available to any individual anywhere within their jurisdiction. An executive or operator should be able to perform their function with ease and familiarity wherever they may be required to work. Information Technology can enable airports to go forward successfully and meet head-on the challenges of the new commercial age.

The challenge to move to value based IT procurement

The challenge for Airport Managers and IT suppliers alike is to approach information management from a value perspective. Value does not mean low price, in fact very rarely does a low price return real value, rather it is the return on investment that may be achieved. By moving from looking at procuring specific systems as cheaply as possible, to implementing information services that will best serve the business objective, the concentration shifts to the engineering and implementation capability of the supplier. Cost becomes a lesser criterion if it is balanced against real, quantifiable benefit.

It will be unusual if the full potential value of an IT system can be realised in a single department and in isolation from others. By its very nature, information enriches and empowers; it knows no boundaries. As all intelligence services recognise, each item of data assumes far greater value and power in combination with others, often in quite unexpected areas.

Where to find value

Given this logic, it is hard to reconcile maintaining information management systems in isolation. The reasons for it, when it occurs, are usually organisational and budgetary. As a result of sound business practices, organisations have profit centred departments with suitable incentives to achieve results. These departments are unlikely to be willing to sacrifice their good figures to provide a contribution to systems that will provide added value to other departments, when a more limited spend will meet the immediate needs of the department concerned. There are also the political considerations of ownership, power and risk to the continued independence of the department.

Therefore, a more global approach should be taken to the evaluation and co-ordination of the information needs of the organisation as a whole. Thus, duplication will be eliminated, or reduced to a minimum, data will be combined to present the most effective information to all departments, and in all probability the organisation's total cost of procurement and management of IT systems will be reduced significantly.

Airports Look to the Future summarised the situation thus:

> There is not a single issue or trend that surfaced in our Airport Study that doesn't depend on the same invisible thread: information. Because airport leaders are beginning to recognize information as the currency of the future, many rightfully place the development of information technology (IT) at the head of their lists of investment priorities.
>
> IT can offer a simple path through complex roadblocks in airport management, significantly impacting the operation's productivity and the customer's experience. "Smart card" technologies can be used extensively in security, ticketless travel, customs, and immigration, all the while providing a stream of information about consumer needs and patterns. Technologies such as relational databases, kiosks, inter/intra/extranets, and data mining can increase an airport's ability to attract customers and identify high-traffic, high-spending consumers for targeted programs and marketing. Just having consumer information presents an advantage to an airport because such information is valuable to retailers, airlines, and others for their own marketing purposes.
>
> Security is an area that's recognized as important to customers, employees, the community, and the cargo industry, yet it remains untapped in terms of revenue generation potential. With thoughtful, reasoned integration, security procedures can capture additional information about customer usage patterns and preferences. This is an especially significant opportunity for airports not yet under construction, where designs and plans should integrate the physical and information environments—who is in the airport, for how long, and for what purpose. For example, how could a high-security alert impact operations—people flow, retail, business services, and flight activity.

Also,

> In the same way, knowing more about your customers will help you capitalize on more opportunities for non-aeronautical revenue—for example, name-brand retailers and special events. The new Hong Kong airport is planning to use its IT infrastructure backbone to capture and market information about passenger traffic and other operations within the airport. Companies will be able to plug directly into the airport systems to access this type of information.
>
> Information and IT can be an airport's most powerful tools. In a competitive environment, providing a competent service, delivering quality, and enhancing the experience beyond that of your competitors will ultimately win the customer. IT's value-added components will

capture and market information about passenger traffic and other operations within the airport. Companies will be able to plug directly into the airport systems to access this type of information.

Information and IT can be an airport's most powerful tools. In a competitive environment, providing a competent service, delivering quality, and enhancing the experience beyond that of your competitors will ultimately win the customer. IT's value-added components will lead to efficiency today and sophistication tomorrow. Although IT itself may not generate revenue, increased business as a result of the application of IT surely will.

Better understanding a new customer. Identifying and capitalizing on new opportunities. Creating mutually beneficial partnerships across the value chain. Airport leaders increasingly understand that meeting challenges like these will depend on relevant, accurate, and timely information. Those same airport leaders are turning to IT as the means to access, analyze, and act on information that will give them the competitive advantage.

The Key findings were:

Inadequate customer ID and tracking is limiting revenue potential on both airside and landside.

> *ACI estimates that as much as 20 % of parking revenue is lost because of inadequate identification and tracking of patrons. This opportunity is especially significant at large hubs where parking accounts for as much as 30 % of non-aeronautical revenue.*

Another example is inadequate tracking of airside cargo handler ground movement, which limits the collection of apron and ramp fees.

A study by BAA and Bechtel estimates that construction costs can be reduced by as much as 30 % through more efficient project management, scheduling, labor, and materials management.

IT can stretch beyond the core functions of airports into value-added areas that will increase non-aeronautical revenue and create new revenue streams. Smart cards will enable airport management to track passenger demographics and understand buying patterns in such a way to increase sales.

Air cargo typically spends 85% to 90% of its delivery time on the ground waiting for information, rather than in transit.

Developing the Future Aviation System

The paper concluded by stating:

Viewing the changing world through the consumer's eyes is the most compelling force behind the transformation of airports. In every consumer segment—from business to leisure, from baby boomer and "X" generations to retirees—there exists a large demand for options, customization, and creative travel experiences. Higher incomes, more flexibility, and a greater thirst for the new and different are more common now than at any other time in history.

Successful companies—from manufacturing to entertainment and all the industries between—are rapidly evolving to take advantage of the new consumer power in the marketplace. Airports are poised to take more control of the travel experience. Great opportunities await the modern airports that meet these challenges by adapting successful techniques and processes from other industries, becoming progressive marketers, and using information and technology to create new travel experiences.

The EDS study convinced us of many things, but this chief among them: It is a great time to be a first-rate airport on the move toward the new millennium.

The author

Mark Sawle Thomas is responsible for co-ordinating EDS' airport business in Europe, the Middle East and Africa. He has spent a lifetime in the air transport industry. After a career in the Royal Air Force as an Air Traffic Control Officer, which included a period as a Data Systems Specialist, he joined the computer services industry as an ATC adviser, moving naturally to work with airport information systems. In his current role with EDS he is part of an international team dedicated to improving information management services to airports.

4 Airport security

Peter Wilkins

Security as a fixed requirement

It has to be accepted from the outset that there is no particular reason to expect that the threat of unlawful interference with civil aviation will be lifted in the foreseeable future. There exists no credible mechanism through which such a cessation of criminal violence could be achieved and be guaranteed to last. Much of the world continues to be politically unstable, and civil air transport is, by its very nature, particularly vulnerable to those who would use it as a target, in one way or another, to achieve their desired objectives. For these reasons, aviation security has therefore undeniably become a major and significant component of airport operation. This has been a basic fact of airport life for many years, and it would be irrational now to assume that the position will improve in any time horizon that could meaningfully be calculated. It could become worse, however, and this could happen overnight.

Another early point which it is appropriate to make is that the words used here in connection with the threat to civil aviation have been chosen with care and deliberation, and for good reasons they make no concessions to the more popular terms used habitually by the media. An act of unlawful interference is a criminal act, whether it involves the seizure of aircraft, or sabotage or other attacks on aircraft, airports or air navigation facilities, regardless of the political persuasion, allegiance, motivation or state of mind of those responsible for it. This blunt truth is important, because it should help to put into proper perspective the measures that have had to be developed and introduced to safeguard civil aviation form unlawful interference. The threat does not constitute a straightforward business risk, to be assessed along with others and included in a sensitivity analysis, because there can be no accurate calculation of the percentage probability, timing or severity of an occurrence. It is simply the case that the civil aviation industry is particularly vulnerable to random instances of unlawful interference, carried out by people whose real objectives are not concerned with the industry but with state interests. This is the situation in which the industry has to operate, a position that is as enduring as, for example, the necessity of carrying out aircraft maintenance.

The term 'aviation security' has also been carefully selected, in preference to the more common 'airport security', a phrase which may represent an all embracing journalistic oversimplification, but which ignores the complex nature of the civil aviation industry, of the government and other agencies involved, and of aviation security itself. Of course, the majority of security measures are carried out at airports, but that is not the issue. Aviation security is concerned with the totality of the measures and procedures that have to be in place and effectively carried out. State governments will make their own decisions on where responsibility for each specific security measure will lie. This is the approach enshrined in the forum which establishes international standards for aviation, the International Civil Aviation Organisation (ICAO). Annex 17 to the Chicago Convention, 'Security', sets out the standards that are expected to be implemented by ICAO's contracting states. How the principles covered by the standards are to be implemented, and by whom, is for the individual states to decide, since they must enact the necessary legislation, and because they have the total responsibility for aviation security within their jurisdiction. Guidance and advice is available, but it is the overall principles of aviation security that are debated and agreed internationally in ICAO.

Reaction to events

It is an unfortunate characteristic of aviation security that the opposition will always have the initiative, and it is therefore particularly difficult to take a positive, as opposed to a reactive, role in the fight against unlawful interference. Historically, the introduction of new security measures has usually followed an increase in the number of incidents, often of a particular type. Thus as acts of seizure became common during the nineteen-seventies, the screening of passengers and their cabin baggage was widely introduced. In 1985, the destruction in flight by sabotage of an Air India wide bodied passenger aircraft resulted in the introduction of hold baggage reconciliation, and triggered research into methods for the screening of hold baggage. The loss by sabotage of two more aircraft, and in particular that of PanAm flight 103 in December 1988, speeded this process, and hold baggage screening is being gradually introduced as suitable equipment is becoming available.

Another distinctive feature of the subject is its political dimension. Following an act of unlawful interference, governments will want to adopt a high profile in closing whatever loophole in the system may have been exploited, to prevent a repetition. They may be criticised for overreacting, but it is difficult to see how else they could behave. The wreckage of Pan American flight 103 fell to the ground in Scotland, where eleven more people were killed in addition to the two hundred and fifty-nine on board, and the flight had departed from the United Kingdom's principle airport. Anyone with access to television or newspapers could not fail to appreciate the full horror of the disaster. It is not surprising, therefore, that, while the effectiveness of that country's security systems was not in question in this tragedy, it is that same country which is in the lead in introducing hold baggage screening.

Attitudes to security

Reference has already been made to the unpredictability of events. An essential difference between security and other areas of airport operation is that the statistics of acts of unlawful interference do not constitute a reliable guide in analysing risk, and therefore are of no help in establishing future requirements. This factor alone, which denies management one of the tools on which it normally relies in the making of decisions, means that security is often misunderstood by those with little or no experience of the industry and of its recent history, or who simply have short memories. For this is not a recent phenomenon, since there have been instances in the past of marked reluctance even by some experienced airport managers to accept that aviation security had become a permanent feature of their work. It is instructive to remember that there were those who, faced with the introduction of new measures which were disruptive of their well established procedures, expressed the view that these security requirements were a phenomenon that would pass away before very long. It was inconceivable, they believed, that acts of unlawful interference would continue indefinitely. The measures introduced would prove effective as a deterrent, it was argued, and in due course those who would mount attacks on civil aviation would in some way lose their motivation, and desist. In short, the problem would go away. Some of the more commercially oriented managers were therefore very unwilling to make long term changes, necessary to raise the standards of security, to existing facilities, and there was a genuine resistance to introducing security considerations as a major design feature of new construction.

This rehearsal of past attitudes is by no means an exaggeration. There have been instances of terminal extensions and new terminals, still in the design stage long after unlawful seizure of aircraft had become common, opening with design features which failed to address certain basic security requirements. It must be said that the regulatory system was also undergoing change in those days, and was less well equipped than now to control project development in a way that would ensure that a new facility met existing security needs, or even anticipated those of the future. Airport operators recognised this only too well, and in extreme cases were prepared to take the line that no government would force them to pull down a new facility once it had been completed, just because it ignored certain requirements.

Security experts all over the world find themselves faced with a challenge to their credibility. They well understand the nature of the threat faced by the industry, but they also have increasingly to recognise that convincing their top management that the threat is permanent may be a different matter. If there have been no recent attacks, management support for the security programme may seem to be diminished. The statistics of events are often enough quoted as an indication that the threat has become less serious, leading to the uninformed conclusion that the security measures currently being implemented therefore represent an over cautious approach. Security, after all, costs money and interferes with the concept of simple and effective airport operation. Such an argument will naturally collapse immediately a single security related tragedy occurs, because in the aftermath it becomes abundantly clear that the recent statistics have proved to be irrelevant. Unfortunately, security, like safety, is often only well regarded by many people when it is too late, and damage has been done.

It may be, then, that unless the lessons of the past are fully learned, industry attitudes may cycle between an acceptance of the need for firm action immediately following an incident, and growing complacency after a period of quiet. This may be compounded if the turnover of very senior management is relatively high, and if what seem to be common company management attitudes are to be maintained even in the highly specialised field of civil aviation. Airports are increasingly being regarded as normal commercial businesses and are expected to perform as such. It is open to judgement whether the usual systems by which businesses are compared and their performance is measured can take full account of the need for aviation security and of the consequences of an attack.

The question of who should pay for aviation security is obviously fundamental to attitudes adopted by commercially oriented businesses in civil aviation. If governments always provided the necessary funds, and shouldered any commercial liability that might be suffered in the introduction of a new measure, then everyone would no doubt happily cooperate in the arrangements for carrying out security requirements in the most effective and efficient manner. As it is, governments are generally reluctant to pass on the cost to their taxpayers, preferring that the user - the ultimate customer, the passenger - should pay for security, being its main beneficiary. Since airports and airlines want to be competitive and must therefore keep their costs down, and because their security activities are carried out at the behest of governments, security tends to be regarded as an imposition, rather than as an integral part of the business. This situation sets security apart, despite the fact that, as with safety, a moral obligation is involved, one on which lives may depend.

Designing for security

This perspective of past and present attitudes is of particularly relevance to any assessment of the future requirements for aviation security at airports. Given that a continuing and evolving threat is something that has to be accepted as a norm, the only practical way forward for the airport operator is to incorporate security needs into every aspect of the design and running of airport facilities, and, through close cooperation with regulatory authorities, to ensure that government measures make security sense when in operation. In this way information and advice that results from government research and intelligence resources can be matched with the expertise and experience of the airport operator in the most practical way.

Good aviation security begins with good design, and this applies just as much to extensions and redevelopment as it does to completely new buildings. ICAO Annex 17 has a standard on this subject, which reads as follows:

'Each Contracting State shall ensure that the architectural and infrastructure-related requirements necessary for the optimum implementation of international civil aviation security measures are integrated into the design and construction of new facilities and alterations to existing facilities'.

It is a basic and very important point, because of the fact that many of the world's major international airport terminals were designed, constructed and commissioned before the first security measures, those against seizure, were introduced in the mid nineteen-seventies, and, even where they have been subject to redevelopment, some primary security requirements have yet to be incorporated. Perhaps the best example of this involves segregation, the arrangement whereby those who have been subjected to security control procedures are prevented from mingling with those who have not, or at least not searched to the same known standard. Outbound passengers, in many older terminals, are still using the same piers to reach their gate rooms as those passengers who are inbound, even though segregation may be adequate in the building itself. Unless the search process is located at each gate, therefore, the departing passengers could, in theory at least, receive weapons or devices from arriving passengers, who may have been subjected to a less stringent standard of search in the country of their departure. While there should be procedures in place to minimise at least the risks inherent in this mingling, they cannot be as effective, either in terms of security or of operating efficiency, as in a fully segregated facility.

Although this is only one of many features of airport design with a significant relevance for security, the example is a good one. It is a common enough situation, and illustrates very well a dilemma which faces airport operators and regulatory authorities alike. The ICAO standard quoted above refers to 'alterations'. At what level of alteration or refurbishment of the system of piers will it become essential for physical segregation to be included in the project? Will the airport operator and the regulator readily agree on when this point has been reached? The airport operator may be reluctant to embark on the capital investment programme needed to replace the old piers until it has to, because the money has to be found to replace a facility which may still be within its original design life. The regulator will want the improvement made as early as possible, but is only too aware that to force the full implementation of segregation on an airport may put a serious strain on the economic viability of a particular facility. Although the effects of operational disruption during reconstruction must be not be underestimated, the main source of the problem will be the financial implications. A balancing factor should be that prior to segregation being achieved, the airport operator will be facing significant expenditure on the alternative, which will involve the use of security staff to carry out searches and patrols, while segregated piers also provide a high level of service to the passengers using them.

It is often been claimed that security must necessarily run counter to the provision of good passenger service, and this may indeed be the situation where the terminal facility predates the introduction of security, and is not suitable for adaptation. In newer buildings, many features of design which take full account of the needs of security will also facilitate passenger flows, and properly laid out facilities are bound to minimise delays in passenger processing. The need for security as a permanent feature of design having been accepted, planners and designers have every opportunity at the earliest conceptual stage of a development to achieve the balance needed to ensure both good security and good service. The architect will have to take account of the priorities, of course. An airport terminal is a functional building, and the aesthetics of aspects of its appearance may have be subordinated to meeting the essential needs of its operation.

The starting point has to be a basic evaluation of the features of the airport's location and its physical layout. Obvious matters which will have to be considered include whether the location is urban or rural, and whether certain associated facilities such as hotels and bus and railway stations are adjacent to the airport complex. The configuration of the roads system that serves the airport is of particular importance, so that the effects of attacks by vehicle bombs can be assessed. It may be, of course, that this type of threat will have to be countered by having a contingency plan, since it is impossible to make an airport proof against such externally mounted attacks. The road access to restricted areas and the layout of screening points are primary considerations, and the possible use of control posts as emergency screening locations, perhaps during the evacuation of a terminal, must be considered. The design of terminal forecourts should facilitate surveillance and make concealment for explosive devices difficult. Car parks, if ideally sited for the maximum convenience of airport users, are often very close to terminal buildings, and the upper floors may overlook operational parts of the airport, providing a line of fire for an attacker. The siting, physical characteristics and surveillance of car parks must therefore receive close attention.

To minimise the effects of an explosion, construction methods and the materials used must be carefully chosen. Extensive research will be necessary, and this will need to be constantly updated as new building methods are developed. The scope of research must range from considerations of the effect of a vehicle bomb on the structure of the building to studies of the results of a smaller explosive or incendiary device on the internal fittings and trim. So far as the structure of the building itself is concerned, the task is to find a method of construction that satisfies current trends and keeps the costs within reasonable bounds while at the same time ensuring that an external explosion will not collapse the entire structure, or cause widespread internal damage and extensive injuries to those inside. For example, while glass is popular for modern construction, its use presents obvious and serious problems. As always, a balanced approach will be necessary, and there must be many options available short of constructing what is in effect a military fortification. Internally, the choice of materials should minimise the risk of secondary damage. The components and container of an explosive device present one kind of hazard, especially if it has been designed to cause maximum injury, while secondary fragmentation, caused by the effects of the explosion on the fittings and fixtures, can be reduced by a combination of careful design and choice of materials.

Attacks have been made in the past on specific groups of passengers or staff, using automatic weapons and grenades, and have taken place in the public areas of terminal buildings, for example in a particular check-in zone. The need to protect people regarded as being at increased risk of attack must be provided for throughout the life of a terminal building, since changes in traffic patterns or political circumstances may introduce the need for special precautions.

It should not need to be emphasised that adequate space must be provided for carrying out the necessary search procedures, whether these involve staff, passengers, cabin baggage or hold baggage. The search facility must be capable of efficient and effective operation even at times of peak flow, and any shortage of space will cause congestion and flight delays, and there may be pressures to avoid this situation by taking operational shortcuts, thereby compromising security.

In planning access control arrangements, the main objective should minimize both the number of points at which controls have to be in place and number of people requiring access. This can be helped by reducing to a minimum of the total size of the area to which restricted access applies, while access to the general 'airside' areas should equally be made as restrictive as possible. Access by authorized people must be controlled, while at the same time unauthorized access must be prevented. Modern automated systems for access control and surveillance can be extremely effective, but they will need to be included in the basic planning of the facility and of its operational procedures. When this is the case, many benefits can be achieved, in the fields of aviation security, the protection of assets, and the management of staff.

Perimeter or restricted zone security inevitably presents considerable problems. As with the case of the structure of a terminal building, it has to be assumed that something well short of a military encampment will be provided, even though aircraft have to be protected from interference while on the ground. The boundary of an airport may be many kilometres long, and surrounded by anything from urban or industrial development to open country. The choice of boundary fence has to cater for the need to avoid undesirable radar reflections, and its height may be constrained by aircraft operational considerations. Whatever its design, it is unlikely to prevent a determined climber from scaling it. If the whole of the perimeter cannot be visible at all times from control posts, electronic means of surveillance may be required, which, if it is to be effective, will need careful design and management. Various forms of automated detection systems are available, but the equipment must be chosen to meet the specific circumstances of the site. It should be emphasised, however that perimeter security alone cannot protect aircraft, facilities and installations from determined attack, and it will be remembered that mortar bombs have been used to attack an airport from a launching site outside the perimeter.

Security equipment

The choice of security equipment and all aspects of its installation and space requirements is a major issue related to the physical design of airport facilities. The most significant new installations will be concerned with the introduction of the screening of all international hold baggage, which is now in the process of implementation or being planned in many states. Just as the measures against seizure were introduced in the wake of an increasing number of incidents of this type, so the destruction of three wide bodied passenger planes by sabotage between 1985 and 1989 has been the trigger for research into methods by which international hold baggage can be screened without causing major disruption to civil aviation. ICAO, in the current edition of Annex 17, has a recommended practice, not a standard, on the subject. *'Each Contracting State should establish measures to ensure that checked baggage is subjected to screening before being placed on board aircraft'*. Screening is defined in the Annex as *'The application of technical or other means which are intended to detect weapons, explosives or other dangerous devices which may be used to commit an act of unlawful interference'*. It is probable that the screening of hold baggage will eventually become a standard, but the next edition of the Annex is unlikely to go that far. The problem is still with the technology, although good progress is being made in that field, and semi-automated equipment is now being brought into use at some airports.

Semi-automation has an advantage over manually operated equipment in that it helps to reduce the risk that human fallibility may result in a failure to detect a device. The work is difficult and repetitive and operators have to be relieved after comparatively short periods if the continuing effectiveness of the system is to be ensured. The current technology involves computer assisted x-ray machines designed to detect a specific component of an explosive device, usually plastic explosive material The aim is for the equipment to make an initial decision without human intervention, using highly trained staff to investigate rejected bags, and to decide whether the original rejection is for a genuine reason, and not a false alarm of some sort.

Screening hold baggage

There are three basic options for the positioning of hold baggage screening operations: before check-in, or 'upstream' as it is known; integrated as part of the check-in procedure; and after check-in, or 'downstream'. The choice will depend on a number of factors, including the existing layout of the relevant terminal facilities and the extent to which these can be modified or extended. In many very busy terminals, the congestion that will result from the location of screening equipment upstream of check-in will be unacceptable on a long term basis, while modifying the check-in desks may not be practicable. The downstream solution will therefore be popular, even though it brings with it a number of complex problems. One is that the baggage make-up area will have to be modified to incorporate the equipment and the work stations for the operators of the system, while still having to cope with the task of sorting baggage and ensuring that it eventually reaches the appropriate aircraft. Another is concerned with the need to locate passengers who have checked-in bags which are subsequently classified as suspect, a problem that does not exist if the screening is 'upstream' of check-in.

It is to be hoped that the development of equipment suitable for screening hold baggage will continue to the point where comparatively small quantities of explosive material can be detected with higher degrees of reliability and lower rates of rejection due to false alarms than are currently available. Whether there will ever be a single machine capable of detecting every possible component of an explosive device to acceptable standards and capable of operation in an airport environment is open to question, but improvements on equipment now being installed can be reasonably expected in due course, and there will always be pressures to replace existing machines with the latest improved technology. This factor could present particular difficulties for airport, and whoever is paying for the equipment. A detection system could have been comparatively recently installed, but a newer and better system could become available, with governments ever anxious, understandably enough, to see it introduced as soon as possible. At worst, given the current preference by governments for the industry to bear the costs of security, airports or airlines could be faced with a constant need to employ the latest technology, regardless of the design life of their existing installations. It is a matter that will no doubt continue to exercise the minds of all concerned.

Who should carry out the work of screening baggage, where this has been is mandated, is something that will be subject to considerable variation. Generally, airports should obviously take the lead in developing and installing the systems to be used, because they will need to offer a screening facility that meets government requirements to the multiplicity of airlines that use their terminals. Precisely who will be responsible for the operation and maintenance of the system, and who may therefore be liable in law if there is a breach of security leading to a disaster, will have to be carefully determined between all the parties concerned.

The need for screening hold baggage has often been questioned by those who have traditionally placed their reliance on forms of profiling, the elimination of the majority of passengers as a threat through the asking of questions, thereby reducing the numbers of bags that have to be examined. Such systems may have been effective in the past, although it is virtually impossible to be certain, but they present a number of problems in the busiest of terminals, causing congestion and being subject to language difficulties. They may, therefore, be better than nothing, but whether they are, in all cases, an effective alternative to the use of modern technology for screening all bags is another matter.

Questions have also been raised about the future need for reconciliation. This process, which seeks to ensure that a passenger who has checked-in a bag subsequently boards the aircraft, is thought by some to be unnecessary if all bags have been screened. It is another example of convenience thinking, and ignores the reality that no single security measure can ever be one hundred per cent effective. Good security will always depend on a series of layers of measures, and it is simplistic to assume that the latest addition can render others superfluous.

The general intention seems to be that hold baggage screening should, where introduced, apply to the bags of international passengers. This would seem to be one of those cases that could be subject to immediate change if an act of sabotage is carried out on a domestic flight. Domestic passengers have been generally included in measures to prevent seizure, and for good reason, since a number of seizures have involved domesic flights. If a group is seeking to damage the interests of a state, and is either based, or capable of operating, in that state, a domestic flight may well represent a suitable target.

Cargo

Cargo may well be regarded as another route for the introduction of an explosive device on to an aircraft, and it may be far more difficult to circumvent than is the case with passengers' baggage. Procedures in the past have been based on the assumption that if the shipper is known to the airline concerned, then the cargo could be regarded as safe to carry. However, the very complex nature of the cargo business makes such assumptions difficult to sustain, and the extension of regulation to all concerned with the making up of freight loads is being introduced. Cargo examination units are not widely available at airports, although, once again, the position could change rapidly following an incident, and there could be a considerable impact on airport cargo facilities.

When examining the nature of the threat to civil aviation from acts of unlawful interference and its impact on airport operation, it is superficially simple to conclude that, since the industry increasingly has to operate commercially, the burden is too great and the solutions are not acceptable. This is the root cause of the hope, pinned on statistics, to which reference was earlier made. If, as has been advocated, the threat is regarded as permanent and security measures are seen as logical and essential, then the incorporation of those measures as a wholly necessary part of airport operation becomes easier to accept, and more practical approaches to the problems can be realistically developed. There does not seem to be an alternative to this if civil aviation is to continue to meet the growing demand for its services. Here can be no arguments in favour of either of the two extremes of cheaper and easier travel, facilitated by low security standards, or of airports and airlines competing for traffic on the basis of their more impressive security arrangements.

Consistency of compliance

Possibly the only way to ensure that, over time, the threat may diminish, is for every state to ensure that at every airport and on every airline the necessary security precautions are effectly implemented. ICAO can only establish the standards, and obtain the agreement of its Contracting States as regards impementation. It does not have the remit or the capability to police compliance with those standards. Nor does any individual state have the authority to act outside the jurisdiction of its government, although there have been ill conceived attempts at this. Ultimately only concerted international action could ever result in any improvement, and if this is seen as difficult to achieve, than the permanence of the threat must be accepted.

So it is realistic to assume that there may always be countries where standards essential to reduce the threat of unlawful interference will not be met. It is for this reason that segregation will continue to be needed, and that states where the standards are high will be at risk of contamination from states where they are not. This proliferation of risk is compounded by the growing popularity of collaborative arrangements between airlines, which can involve commercial links between those registered in both categories of state. It may be impossible to ensure that, in every case, the continuity of essential security practices can be maintained. Anyone who travels extensively will know of cases where there has been inadequate security at a departing airport, and will therefore be aware of the risk transferred to the arriving airport, as well as to the flight itself. In such circumstances airlines from states where security has a high priority are expected by most regulatory authorities to make good any deficiencies that have been observed. Where an arriving aircraft is part of a series of legs flown by other airlines, as part of a code sharing agreement, for example, the position may be less clear, because of problems of jurisdiction among others.

As will always be the case, the strength of the security chain in civil aviation is that of its weakest link. Only through concerted efforts, by governments and by every part of the industry, can there ever be any hope of improvement. It is unfortunate that greater comfort cannot be offered to airport operators or to the rest of the industry in looking to the future, but, as been shown, part of the remedy at least is in their hands.

The author

Peter Wilkins is an independent consultant on aviation security and aviation-technical aspects of airport operation. Until April 1995 he was Director, Technical/Safety and Security with the Airports Council International at its world headquarters in Geneva, and before that he was with the British Airports Authority, latterly BAA plc., retiring in 1991 from the appointment of Director, Safety and Security. While working for both these organisations he represented the international airports community on security issues at the International Aviation Organisation, having served on the ICAO Aviation Security Panel since it was established.

His aviation credentials go back to the first twenty years of his working life which he spent flying in the Royal Navy's Fleet Arm, latterly as a specialist in airborne early warning. After leaving that service he worked for a time in the steel industry and then for an organisation concerned with the political and financial relationships between the UK public sector corporations and the Government.

He is a member of the Royal Aeronautical Society, a member of the Chartered Institute of Transport and a Fellow of the Institute of Management. Married, with two adult sons and two granddaughters, he lives and works in Brighton on the south coast of England.

Part 2

HUMAN FACTORS AND TRAINING

5 Human factors in the cockpit

Margaret T. Shaffer

Attribution of the probable causes of air accidents to 'Human Error' is quite high-depending upon the source, the numbers are 75 % or more. If one were to include in that number, the human error of aircraft construction workers and maintenance personnel, the percentage would be even higher. Since the attribution of air accidents to human error in the cockpit is so high, it is incumbent upon everyone involved in the aviation community to actively consider the wide variety of human factors issues that exist in the cockpit today. In addition, it is important to consider how those factors are likely to affect the cockpits of the future.

It should be stated at the onset, that human factors issues in the future cockpit are only part of the paradigm. The air traffic control system interacts with the cockpit system to create an environment for safe flight. Either, viewed in isolation of the other, is addressing only half of the picture. Figure 5.1 depicts the relationship between the pilots and air traffic controllers and the 'behind-the-scene' humans that impact the safety of air flight. As figure 5.1 shows, the only factor in the Ground Air Paradigm (GAP) that does not focus on the human in the system is the weather. The task for human factors professionals is to bridge the GAP and provide an appropriate link between the Ground and the Air.

Therefore, although the focus of this chapter is on human factors in the cockpit, many of the issues raised are equally important in the Air Traffic Control system and in the interactions between the two.

Cultural factors

One of the most critical issues which spans the cockpit-air traffic control (ATC) interactions is that of language and culture. As the world airways continue to expand into the developing world, the issues of language and culture both within the cockpit and between the cockpit and ATC become

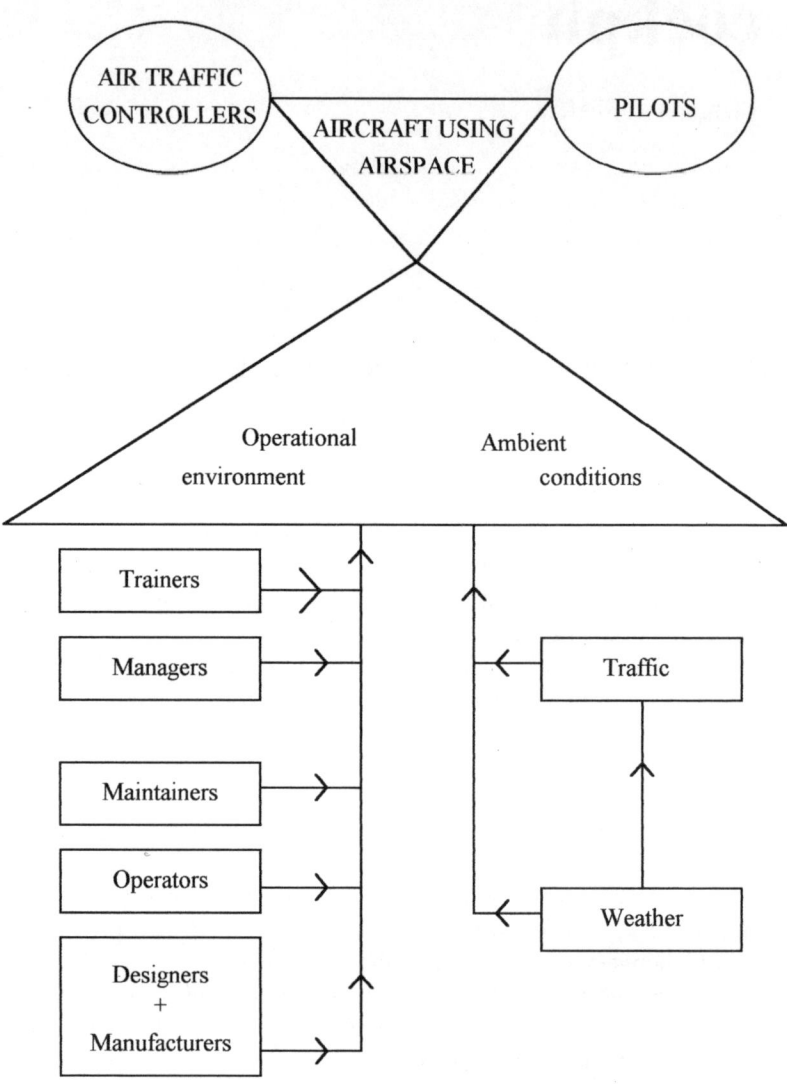

Figure 5.1 Ground-Air Paradigm (GAP)

more critical. English is usually the official language of the interactions between ATC and the cockpit. However, the degree of facility with the language, and variations in accents and dialect make the issue of language and understandability critical.

Digital transmission of some types of communications, for example Advanced Terminal Information System (ATIS) messages, reduces the effect of language transmission for those who have the digital capability. However, the very wide variety of aircraft operating worldwide and the variations in the amount and type of equipment on board are enormous. Many aircraft not only do not have equipment capable of dealing with digital messages, but they may have antiquated radio transmission systems as well. Therefore, the transmission and the understandability of voice communications remain a challenge.

Added to the difficulty in actually understanding what is being said are the cultural interventions of how the message is given. Although there are conventions that guide the construction and content of messages, there is some licence for variations in those messages. The greater the variation in content and format permitted, the greater the opportunity for misunderstanding.

Two other cultural factors complicate communications: the social context of authority within the cockpit and culturally influenced responses to criticism. An autocratic captain can stifle the flow of information within the cockpit. Although the affects of an autocratic captain can be felt in all cultures, it is more prevalent in some cultures than others. Add to this the cultural and organizational factors that influence the acceptability or perceived acceptability of admissibility to making mistakes, both in the cockpit and on the ground, and the issue becomes even more complex.

In *Cultures and Organizations* (aptly subtitled 'Intercultural Cooperation and its Importance for Survival'), Geert Hofstede identifies that there are systematic national cultural differences with regard to values about power and inequality, with regard to (1) the relationship between the individual and the group, (2) the social roles expected from men or women, (3) ways of dealing with the uncertainties in life, and (4) whether one is mainly preoccupied with the future or with the past and present. Furthermore, superimposed on the national cultural differences are organizational cultural differences that affect both business and governments and their respective demands on the individual. The message is clear; all of these factors interact to create an environment in which cultural differences are wide and must be considered. Training must not be limited to technical issues; it must deal with the cultural issues that predispose actions and interactions, particularly when the human is under stress.

Merritt and Ratwatte (1997) address cultural issues specifically within the cockpit. They contend that even within a given country, there is no truly single culture cockpit, that cultural influences exist in all cockpits -- that their influences are a matter of degree. After presenting a debate on the range of cultural milieu that can be present in the cockpit they conclude, 'culture can be a problem in the cockpit, but only if its influence goes unacknowledged'. They continue, 'in any airline, it is management's role not only to define and then recruit the best pilots, but also to diagnose the strengths and weaknesses of the operation, and design organizationally-appropriate, culturally-sensitive interventions which will optimize the safety and performance of all its employees.'

They see several ways in which a multi-cultural airline can promote safety:

- Set clear organizational standards and policies that eliminate previous cultural assumptions and explicate all the organization's goals and procedures.

- Provide adequate training as a tool for enhancing safety. This training should include standards and practices, emergency procedures, CRM, intercultural awareness. For native English speaking pilots, the training should also include effective communications using simple, slow, and precise messages.

- Provide recruitment of pilots which includes not only consideration of their technical competence but also their potential to be interculturally effective.

One might add an additional critical element to this list of interventions aimed at reducing any negative affects of cultural diversity within the cockpit and between the cockpit and ATC. *It is essential that they be applied worldwide and with the same level of commitment in all countries.* While difficult, it is a goal that the aviation community should strive to achieve. Anything short of this could severely compromise safety in the skies of the future.

Cockpit automation

Much has been written regarding the impact that automation has on the modern cockpit. Clearly, cockpit automation is and should be here to stay. The question is how much and what kind of automation is most effective in providing both safety of flight and effective piloting? Charles Billings (1995) talks about Human Centered Automation, in which he stresses the importance of keeping the pilot in the loop, and in a state of situational awareness throughout the flight. The controversy stems from the question 'what is the correct type and amount of automation'? Automation should not be designed based solely upon what it is easy to automate, leaving the human to perform the 'leftovers' i.e., those things that are more difficult to automate. Literature on job design has long shown that productivity and effective job performance are dependent in part upon the human having a meaningful interaction with the automation. Furthermore, it is important that the human use his or her judgement when interacting with automation. There are inherent dangers in implicit trusting of automated systems, with the human not questioning when the automated system provides information or performs functions that are counter to the intuitive judgement of the pilots. Under those conditions, it is incumbent upon the human to draw upon the collective experience present in the cockpit, information available from other systems in the cockpit, and perhaps information available from ATC or other aircraft in the area to override the automation if necessary. It is at this point that effective cockpit resource management becomes a critical element in the management of the automation. Although Billings does not provide specific guidance on what to automate or how to automate, he is clear that the human must be the central focus and that the human must be not only 'in the loop' but must be central to the loop.

Keeping the human in the loop is not a trivial matter. Several problems occur when automated systems prevail. First, when automated systems are operating, the human task is to monitor their activity. Luczak (in Salvendy, 1997) indicates that a human operator is monitoring a system when he/she scans an array of displayed information without taking any action to change the system state. This is performed to update his knowledge for an appropriate decision preparation. Decision-making implies a selection of an appropriate alternative from a set of possibilities, based on sources of information about the system state. As such, system monitoring or vigilance is a task that humans are notoriously poor at performing. Boredom is counter to effective attention. Add fatigue and its impact on effective attention and the problem becomes even more complex. Even with checklists and established and practiced scan patterns, the task of monitoring is capable of lulling even the most conscientious pilot into complacency.

Flight status and mode indicators become more important as automation is increased. The use of multi-media presentation modes offer some opportunities to capitalize on new and existing technologies to improve the effectiveness of the pilot in the loop. As future automated systems are developed, and automation increases it will be even more important to consider and build in strategies to keep the pilot in the loop and involved in the piloting task.

The following are adapted from Chapanis (1996) who provides some strategies for appropriately capturing attention and keeping the pilot in the loop.

- Alarms should be compelling under all operating conditions, and must be immediately recognizable as alarms.

- Each abnormal condition should trigger a unique alarm. It should be easy to determine the source of the problem.

- The alarm should convey information about the approximate level of danger, hazard, or abnormality In the case of multiple simultaneous abnormal conditions alarms should be prioritized according to their importance or degree.

He also identifies an important difference between sensing and perceiving. Sensing is the simple reception of stimuli. Perception, however is the temporal or spatial organization of simple sensory information into meaningful wholes. For effective monitoring to take place, e.g., for an appropriate understanding of the significance of an annunciator alarm, the task is a perceptual task. As such, it requires appropriate learning on the part of the operator. In addition, he indicates that more attention should be paid to the design of the things to be perceived.

Although these guidelines have been discussed as part of improving monitoring and increasing situational awareness under somewhat normal automated conditions, they are even more important when associated with the occurrence of an emergency situation. These may be situations that the automation is not equipped to handle or the situation within which some aspect of the automation itself is nonfunctional.

In most countries, there are very strict regulations about currency and proficiency, and check rides are required as a regular part of keeping pilots current. During these check rides, emergency procedures must be performed. However there is a big difference between knowing what one is supposed to do in an emergency and being able to perform it effectively in an actual emergency. Functions that were routine when performed frequently each day are not so routine with automation when they are practiced only as a part of 'emergency procedures'.

Training becomes a critical part of this equation. The performance of some tasks may require 'over-learning' so that they are reduced to automatic responses to ensure that they can be performed under the most extreme adverse circumstance. Again, it is critical that *everyone* who is permitted to pilot the aircraft be adequately trained to handle these emergencies. As important as good training is, it must not be considered a solution to correct poor design.

Traditionally new systems are tested by test pilots whose training is often different from the ordinary line pilot. Although they may not be 'better' pilots, they are trained to interact with new systems and to critically evaluate the system performance. As such, their performance with new systems may not be typical of what might be expected when typical line pilots interact with the new system. The potential problems occur when those pilots who are the least qualified but who nevertheless are certified to fly attempt to use the systems. The presentation of information and the design of control mechanisms must be oriented to the abilities of the least qualified. As systems become more and more complex, the importance of this issue increases in magnitude and the need for common certification standards become more critical.

The U.S. FAA Human Factors Team (1996) made five recommendations relative to the management of automation in the cockpit. The following are extracted from those recommendations. They state that the FAA should:

- Ensure that a uniform set of information regarding the manufacturers' and operators' automation philosophies is explicitly conveyed to flight crews.

- Require operators' manuals and initial/recurrent qualification programs to provide clear and concise guidance on autopilot and auto throttle operation and circumstances, under which they will engage, disengage or change mode.

- Review of the autopilots on all transport category airplanes to identify the potential for producing hazardous states or undesirable manoeuvres. Results of this review should be the basis for initiating appropriate actions such as design improvements, flight manual revisions, additional operating limitations or changes in training or operational procedures.

- Conduct analyses to better understand why flight crews deviate from procedures, especially when the procedural deviations contribute to causing or preventing an accident or incident.

- Request industry to take the lead in developing design guidelines for the next generation of flight management systems.

These recommendations, when implemented, could greatly contribute to better human/automation interaction and increased safety. It is critical, however, that they be applied universally worldwide.

In today's cockpit the ability to provide information is almost limitless. Ask most crew members and they will tell you that they want as much information as possible. The philosophy is 'more information is better'. However, that philosophy should be re-visited. More isn't necessarily better. There is a point of diminishing returns where more information contributes to information overload. Now that flat panel

Developing the Future Aviation System

displays permit the display of vast arrays of information in a wide variety of formats, the issue of prioritization of information becomes more important. What information is needed, and when is it needed? When information is not needed, how may the crew member declutter his/her display so only essential information is presented? And, how can that crew member re-acquire instantaneously that previously decluttered information if or when it becomes critical? Again, it is critical that the informational requirements of *all* pilots be considered, and that the presentation of this information be prioritized.

Not only must the information be prioritized; it must be appropriately presented. In a recent study for the US Air Force, Shaffer et al found the following generic attributes to be most problematic in a new upgraded military cargo cockpit display:

- Visual accessibility

- Location and arrangement of components

- Ease and accuracy of control

- Legibility

- Adequacy of information

- Menu Structure

- Use of Colour

- Size and shape of display elements.

These generic attributes were independent of the specific display system or subsystem and were found to be the most significant for all systems and subsystems. The selection and presentation of information is central to all of these elements except ease and accuracy of control. Each of these factors in itself is an important human factors issue and must be addressed with the most up-to-date data possible. Therefore, it appears that in our design of new aircraft systems, we need to pay particular attention to these display attributes and ensure that they adhere to the most stringent human factors guidelines possible.

The interaction of these factors poses an even more formidable challenge. The literature abounds with studies on legibility, menus, colour, size and shape of display elements, visual accessibility, etc. However, it is very difficult to find *non-proprietary* studies that address the interactions of these factors in a systematic fashion. Major aircraft companies have addressed these issues, and the findings are reflected in their cockpit designs. However, it appears that each time an aircraft cockpit is to be designed or re-designed, their designers must 're-invent the wheel'. While it is understandable that the proprietary interests of commercial companies must be maintained, it is important that any information that can be shared with the aviation community 'at large' should be. The greater goals of increasing aviation safety will be then better served.

One published recent attempt to investigate the integration of all aspects of display characteristics on a flat panel display was performed by Toms, et al. (1995) at Wright Laboratories in Ohio. The question being asked was 'is the performance with the new display as good or better than performance with the existing display?' Since existing displays were being utilized by crews in the cockpit, this was considered to 'validate' their usability. However, nothing was said about optimizing performance, or whether performance might have been even better using a slightly different configuration. While its scope was limited to the study of only one new configuration, it was important since it is one of the few recent studies of display integration that is available in the published literature.

Integration of off-the-shelf components

In recent years there has been a greater tendency to use off-the-shelf components in new cockpit systems design. This trend, particularly prevalent in the US military has been driven by both time and cost expediencies. The time and cost required to design systems (or sub-systems) specifically for a particular application is formidable. It is considered more cost effective to try to use a variety of special purpose sub-systems that exist than to design the complete system from the beginning. Each of these sub-systems to be used has been designed and engineered to meet exacting requirements for a specific application. The challenge for the human factors professional is the *integration* of these 'boxes' in *new* contexts.

When a number of sub-systems designed by different companies or organizations are combined to function as a whole system, a number of potential problems can occur. The ability of computers to 'talk to each other' using different computer languages can be problematic. More critical from a human factors viewpoint, is that the methods of interacting with the various sub-systems can often have different operational conventions, and those sub-systems could be co-located. This complicates the human/system interaction and could contribute to an incident or an accident. Furthermore, the sub-systems when linked together may operate in a non-intuitive way.

Developing the Future Aviation System

In the FAA Human Factors Team (1996) evaluation, they cited that a major problem stemmed from system operation that was counter to what was expected. It is critical, therefore for the human factors professional to thoroughly analyze the arrangement of sub-systems, their operational characteristics, and their interaction with the other sub-systems, to ensure an optimized user interface.

References

George, E.J., McCann, C., Shaffer, M.T., and Bolstridge, L.D. (1998), *C-141B All Weather Flight Control System Human Engineering Control and Display Evaluation Final Report: Results from all Phases of Testing.* AFFTC-TLC 98-01, Edwards Air Force Base, California.

Hofstede, G. (1997), *Cultures and Organizations, Software of the Mind,* McGraw Hill, New York.

Chapanis, A. (1996), *Human Factors in Systems Engineering,* John Wiley & Sons, New York.

Billings, C. (1996), *Human-Centered Aviation Automation: Principles and Guidelines,* NASA Technical Memorandum 11038, Moffett Field, California, NASA Ames Research Center.

Merritt, A. and Ratwatte, S. (1997) *Who are You Calling a Safety Threat?! A Debate on Safety in Mono-vs-Mixed-Culture Cockpits,* Ninth International Symposium on Aviation Psychology, Columbus OH, USA.

Luczak, H. (1997) Task Analysis in Salvendy G., (ed.) *Handbook of Human Factors and Ergonomics,* Wiley Interscience, New York.

Federal Aviation Administration Human Factors Team (1996), *Report on The Interfaces between Flightcrews and Modern Flight Deck Systems,* FAA, Washington, District of Columbia.

Toms, M.I., Cone, S., Grier, G., Boucek, C., Brown, T., and Patsek, M. (1995), *Full Mission Evaluation in C-141 Upgrade program.* Volume II, for Flight Dynamics Directorate, Wright Laboratory, Wright Patterson Air Force Base, Ohio, contract no F33615-93-D-3800.

The author

Margaret T. Shaffer, M.A. (Experimental Psychology), B.A. (Music Education and Psychology) is a member of several professional associations concerned with Human Factors and Ergonomics.

She is the President of Paradigm International, a consultancy company, but in addition to her administrative and management responsibilities, Ms. Shaffer has functioned as the technical project director of over 90 multidisciplinary studies of applied human factors, workload assessment, human performance, Manprint evaluation, functional analysis, training, and evaluation in government and industrial settings both nationally and internationally. Her work has included the development of several innovative methodologies for the study, evaluation, and analysis of human/system interaction.

Ms. Shaffer developed an empirically based technique and software analysis program for task analysis and workload assessment through the analysis of videotaped interactions between operators and systems. The analysis provides a baseline for the frequency of performance and the time required to perform the various tasks and communications in existing aircraft missions. Further, it identifies tasks which can be performed simultaneously and which cannot. The latter tasks have been identified as candidates for automation in new aircraft designs. This technique was used in studies performed for the U.S. Army, the Canadian Department of National Defence, the U.S. Marine Corps and the U.S. Navy.

Ms. Shaffer was also instrumental in establishing the research activities in applying psychological techniques and principles to the fields of community planning and transportation planning in the Washington D.C. area. Responsible for the procurement and project direction of a number of studies incorporating a multidisciplinary team approach to problems of housing, transportation, health, welfare, recreation, education, and land use.

6 Laws for the design of the Universal Cockpit displays

Lawrence E. Tannas Jr.

Introduction

The new colour flat-panel active-matrix liquid-crystal display (AMLCD) has created an opportunity for designing the most integrated, functional, and universal set of cockpit instruments in the history of aviation. With the coupling of this development and the evolution in electronics, the industry has truly reached a new paradigm. Examples of the new cockpit in the new paradigm are evolving and are exhibited in the Boeing 777, Lockheed 141B upgrade, Lear 45, Raytheon T-6A, and others—but we are not there yet. These four examples are all new cockpits or upgrades that use AMLCDs and depart significantly in their portrayal of the primary set of instruments.

The thesis is that the industry could and should converge to a single, uniform set of cockpit instruments based on the historical evolution of instrument formats that are infused with flat-panel displays and optimized for situational awareness. The set of cockpit instruments would include a subset for primary flight, navigation, and engine instruments. The discussion herein will concentrate on the primary flight instruments as an example.

The primary flight instruments in the various new AMLCD cockpits do not look the same to the pilot. Levels of integration vary from almost none to nearly complete integration. The functionality is different, with mixtures of tight and loose scan patterns, fly-to and fly-from command needles and round dials and tapes. There is less universality in the present mix of the AMLCD panels than there was before the glass cockpit, as exhibited in a sampling of new AMLCD panels.

The flat-panel version of the glass cockpit offers a significant improvement in design opportunity over the CRT because of the CRT display volume issue. The CRT served the cockpit well and forged the way to integration of the primary set of instruments. The flat-panel AMLCD volume has allowed an integration in packaging, eliminating other boxes and extra cabling, and allowed sensors to be included in the same box. The AMLCD volume enables the concept of a single instrument box, again with electronics, symbol generators, and sensors all self-contained. The single box concept is very important for universality with regard to aircraft size, maintenance, and situational awareness.

The integrated, functional, universal primary flight instrument in the new paradigm is transparent to the electronic display technology. Obviously, the display must have all the performance and features required for aviation. So far, the only display technology that meets all the requirements of the new paradigm is the colour active-matrix liquid-crystal (AMLCD) for the integrated primary displays. At some time in the future a new technology may arrive, and when it does, it could be substituted for the AMLCD. The pilot need not know the difference, and new training should not be required.

The latest AMLCDs have performance parameter capabilities that exceed those of CRTs. The main advantages are in weight, power, volume, and reliability. However, the AMLCD performance exceeds that of CRTs in other areas such as:

- Luminance uniformity

- Resolution uniformity

- Immunity to ambient illumination washout and colour de-saturation

- Fault tolerance

- Night vision goggle compatibility

- Brightness and dimmability

- Increased usable display area to panel area

- Environmental and mechanical qualifications

The CRT has an inherent wider viewing angle and wider operating temperature range than does the AMLCD, but the AMLCD meets the needs of aviation.

Cockpit display objective

The first thesis is that the art and science of flying is transparent to and independent of the vehicle type and the display technology. This is the fundamental guiding principle behind the laws for cockpit design that follow. The mission may be different, the size may vary, the power plant may change, and the configuration may be variable, but the fundamental pilotage tasks are the same. The pilot must be able to cause the aircraft to fly straight and level; maintain a heading, altitude, and air speed; climb, descend, and turn; control power in the presence of turbulence and phugoid and aerodynamic oscillations; navigate; and determine failures, all in meteorological instrument conditions (MIC) without visual reference to the ground.

The second thesis is that the art and science of all flying can be done with a single universal set of instruments, the main sets being for primary flight, for navigation, and for power. The three would give sufficient situational awareness so that the pilot could complete the mission or alternate mission within the full design envelope of MIC.

This quest for the single universal cockpit display set is the objective. It is postulated that the state-of-the-art in aerodynamics, power plants, electronics, sensors, and electronic displays has matured enough to achieve this objective during the transition to flat-panel displays.

However, the new flat-panel display cockpits are not converging into a single universal cockpit but diverging. We can design into the electronic display any visual image desired. In the first step, as in the Boeing 757 and 767, the electronic image aped the classical electromechanical flight director. In the second step true fusion of the classical 'T' instrument set, or primary flight set, was integrated and portrayed on a single electronic display, as in the Airbus 320 and the McDonnell-Douglas MD11.

As this integrated display of the primary flight instruments is implemented with flat-panel AMLCDs in different aircraft such as the Boeing 777, Lockheed C-141B upgrade, Lear 45, and Raytheon T-61, significant liberties are being taken, and the opportunity for a single universal primary flight instrument is being significantly diluted. For example, the speed tapes on the Boeing 777 have the highest speed indicated at the top, while on the Lockheed C-141B (upgrade) the highest speed is indicated at the bottom. The new trainer (JPATS) Raytheon T-6A has no integration and has round dials for velocity and altitude. There are numerous other symbol and colour-coding differences as well.

Laws for cockpit instrument display design

The single universal cockpit instrument design has not yet evolved. To cause the single universal cockpit to occur, it is necessary and appropriate to promote a set of laws for its design. These laws are believed to be invariant and transparent to the display technology.

The first five laws are paramount and relate to socioeconomic issues. The sixth through twelfth laws relate to display images and how they are designed.

Laws for Aircraft Instruments

First Law—Economic Reality. Avionic display technology cannot move ahead of commercial display industry technology. Due to the small production volume and long product life of aircraft, the industry cannot support the infrastructure as a display technology innovator. It is far more economical for aviation displays to be derivatives of commercial displays than to be developed and manufactured solely for aviation.

Second Law—Pilot Culture. Avionic display and control symbology and procedures must evolve from the present pilot community culture. The transition of pilots between new and old aircraft panels within a fleet, or from employer to employer, must be as easy as possible to minimize the necessity of retraining and to enhance safety. Detraining and retraining pilots is to be avoided due to cost and subtle safety issues.

Third Law—Cost-effectiveness. Any changes to the cockpit must be cost-effective. Changed hardware or software must be better and less expensive. New concepts in aviation displays technology must improve reliability and safety and be less expensive. All cost and savings elements include the expense of retraining pilots.

Fourth Law—International Occupation. Piloting is an international occupation, and all changes must converge to common pilot-interchangeable display images and procedures. Every effort that is feasible should be made to make all cockpits as identical as possible. This is done to enhance safety as pilots transition from one type of aircraft to another, from student to pilot, from old aircraft to new aircraft, from carrier to carrier, from military to civilian, from country to country, etc.

Fifth Law—Military Exception to the First Laws. Everything technically and financially possible must be done to achieve and maintain air superiority. Doing too much may cause financial disaster.

Sixth Law—All and Only Data Display. For primary flight instruments, all and only data used for piloting and situation awareness must be updated and be readable at all times.

Seventh Law—Colour as Aid Only. Availability and readability of alphanumeric and symbolic images is transparent to and independent of colour. Colour is orthogonal information to alphanumerics and symbols.

Developing the Future Aviation System

Eighth Law—Failure Recognition and Recovery. Pilot is to analyze and deduce system and/or display failures, and reconfigure and complete the mission objective with single-point failure and generic failure.

Ninth Law—Fly-To. All command symbols and error symbology are to be of the fly-to polarity.

Tenth Law—Inside-Out and Heading Up. All graphic images are to portray polarity as if the pilot were inside the aircraft looking out. Rotating images of maps, weather, etc., are to be oriented with aircraft heading up at the top of the display.

Eleventh Law—Round Scales and Linear Scales. A tape or linear scale is to be used for bounded values and a round dial for unbounded values. Air speed and altitude are examples of bounded values. Heading and roll angle are examples of unbounded values.

Twelfth Law—Presentation of Image. A digital image is to be used if the value is used for reading and an analog presentation if the value is used for flying. Altitude display requires digital images to read altitude and analog presentation to fly the aircraft.

Summary

The single universal cockpit will never occur without the use of common design rules or laws for the new generation of cockpits. The objective cannot be legislated into existence; it can only evolve. Once it has been attained for new cockpits, it may not be cost-effective to go back immediately and change all the old cockpits. Instead, they will eventually be upgraded.

Now that we can make any display image we desire in glass electronic displays, a concentrated effort is needed to cause them to converge to the ultimate single universal display images. However, the ultimate display images are not yet known or, at least if known, not agreed upon.

The air speed is displayed as a vertical tape in most new AMLCD glass cockpits—as is appropriate—with the present speed displayed digitally. This is consistent with the eleventh and twelfth laws. However, to follow the fly-to polarity of the ninth law, with aircraft pitch control the air speed should have highest speed at the bottom of the tape.

The design rules or laws used to design the ultimate display images are postulated herein. These may also change, and new ones may be added before the ultimate single universal display images are found. Once the ultimate cockpit images are found, they will hopefully evolve further in time to improve continuously the art and science of pilotage and the safety of flight. Without agreement on fundamental design rules that are taken as laws, the cockpit display images of the new paradigm will diverge, as they appear to be doing now.

References

Boyd, S.P. (1997), 'Flight deck improvements vs. Commonality: Human factors implications for mixed fleet operations', Ninth International Symposium on Aviation Psychology, Columbus, Ohio, USA.

Coombs, L.F.E. (1990), *The aircraft cockpit*, Butler & Tanner, Ltd., Frome, Somerset, U.K.

McCartney, R.I.; Haim, E.; and Kucera, C. (1996), 'Performance testing of the primary flight instruments for the Boeing 777 airplane', *SPIE Cockpit Display III, V 2734*, April 10–11, 1996, Orlando, Florida, pp. 86-93.

Rupp, J.A. 1986, 'Color flat-panel displays in the commercial airplane flight deck', Japan Display '86 IDRC, Sept. 30-Oct. 2, 1986, Tokyo, p. 406.

Tannas, Jr., L. E. (1994), 'World status of avionic AMLCD panels', *SPIE Cockpit Displays, V 2219*, April 7–8, 1994, Orlando, Florida, pp. 202-212.

Tannas, Jr., L.E. (1998), 'Laws for the design of the universal cockpit displays', *SPIE Cockpit Displays V*, April 15, 1998, Orlando, Florida.

The author

Lawrence E. Tannas, Jr., is President of Tannas Electronics, Orange, California, USA.

7 Creating a culture of safety

Dianne Hill

Introduction

Accident investigations point to flight crew error as the prime factor in approximately seventy percent of worldwide commercial jet fleet accidents. This is not new news. In fact, aviation has spent more than fifteen years studying, creating, and delivering Crew Resource Management (CRM) training. Why? Effective CRM skills lead to better use of resources on the flight deck, ergo safety. When the FAA mandates CRM not only for U.S. airline carriers, but also those carriers arriving in the U.S., there is significant support for the training and the belief that it works. But are we doing the right thing or doing things right?

As a way of addressing the need for increased safety, Crew Resource Management has evolved over the last fifteen years as a way to reduce flight crew error. However, actual CRM programs vary significantly in *duration* (from 4 hours to four days), *content* (accident/incident debriefs to instructor-led assessments of CRM skills displayed in the simulator), *goal* (such as FAA regulatory compliance designed for a more holistic and ambitious change in flight deck culture), and learning methodology (lecture, experiential, individual or team-based). As a strategy to affect the aviation accident statistics, CRM's focus on the flight deck falls short. Neither CRM training nor the flight deck is an island unto itself. Behaviour on the flight deck is guided silently by not only the norms of the organisation but of the industry and country as well. These silent and powerful ways of doing things emerge as culture.

Resource management

Aviation technology as a resource creates an interesting dilemma. At what point is it a resource and at what point is it a liability? To the extent that the

Creating a culture of safety

flight crew understands the technology and uses it effectively, it is a resource. Important to the resource mix are various components, not the least of which is development of the human/human interface in addition to the human/machine interface. How do these components come together in an operational setting? Desired results, R_3, follow the effective use of resources, R_1, through relationships, R_2, within the larger framework of reflection, R_4, or critique, as illustrated in Figure 7.1. For purposes of this discussion, the human-to-human interaction and the potential synergistic relationship characterise R_2 as noted in Figure 7.1. Observation of behaviours in operational settings with new understanding creates reflection, the porous circles depicted by R_4. Reflection interprets data from both internal and external influences, including human and machine.

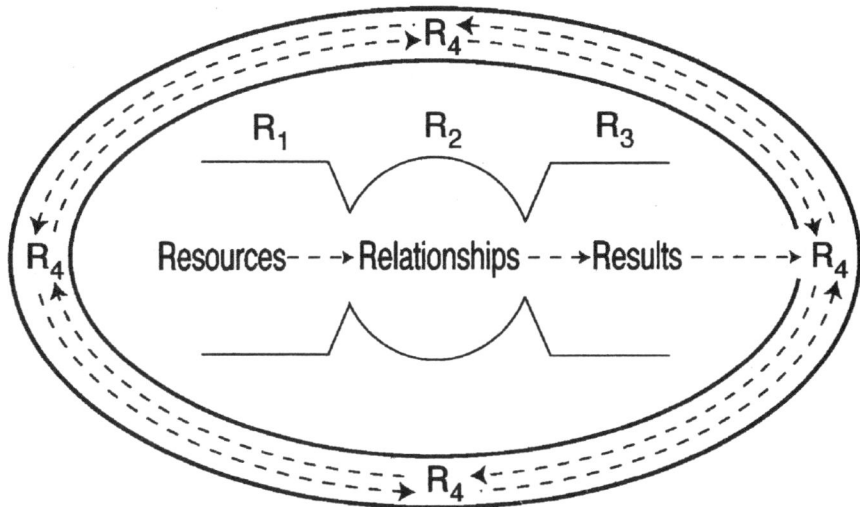

Figure 7.1 The R1234® concept -four Rs of leadership
Copyright © Scientific Methods, 1998 from Grid theory developed by Drs. Robert R. Blake and Jane S. Mouton, 1968.

In the automated flight deck, programs were initially developed to simplify operations, provide more information, and ease overall workload. In operational use, the computer-driven flight management system (integrated avionics, autopilot, and engine functions) actually may have increased complexity for the human factor. In fact, the most often asked question directed toward the flight management system (FMS) today is 'What is it doing now!?' The FMS is indeed a new crew member whose intentions are often not readily apparent. This new crew member must be queried, and if not readily understood, then reprogrammed or rejected in favour of manual

Table 7.1
Decision-making criteria: effective use of human resources
Copyright® Scientific Methods, 1998 from Grid theory developed by Drs. Robert R. Blake and Jane S. Mouton, 1968.

Ask:	If the answer is:	Then use
Who owns the problem?	Only one person	One-alone
	Two or more people	One-to-one or one-to-some
	The entire team	One-to-all
Do I have time to involve the others?	No time to involve others	One-alone
	Some time but not a lot	One-to-one or one-to-some
	Enough time to involve all sources	One-to-all
Do I have the competency to make the decision alone?	A single person has all needed skills	One-alone
	No single person has all needed skills	One-to-one or one-to-some
	Skills of some members will be needed	One-to-all
Do I need the commitment and involvement of others?	No	One-alone
	Yes, to a moderate degree	One-to-one or one-to-some
	Yes, to a high degree	One-to-all
What is the impact on the rest of the team?	Impact is very low	One-to-one
	Impact is low to moderate	One-to-one or one-to-some
	Impact is moderate to high	One-to-all
Is synergy possible?	No synergistic opportunity	One-alone
	Synergy is possible	One-to-one or one-to-some
	Synergy is highly probable	One-to-all
Is there development potential for others?	No	One-alone
	Possibly, for some	One-to-one or one-to-some
	Yes, for all team members	One-to-all

systems. Therefore, understanding among crew members as to what the other is intending and why is critical to challenging both the personal and group autopilot referred to by Geert Hofstede as 'software of the mind'.

Crew culture

Recent discussions surrounding perceived erosion of Captain's authority calls into question not CRM itself, but the lack of understanding of effective CRM principles. When confusion exists as to who participates in what decisions and when, there is a lack of clarity as to where the decision responsibility resides. The final authority resides with the Captain. As illustrated in table 1, a good decision does not mean everyone is involved in every decision all the time. Effective decisions result from behaviour guided by sound assumptions.

Unquestioned obedience to leadership, as well as low regard for leadership, restricts the open flow of information and critique required for safe flight. This can present a challenge to the effective use of resources involving inter-cultural relationships. Captain Yamamori of Japan Airlines commented that 'CRM is culture free'. An aircraft knows neither who's flying it nor the crew's country or company culture. An aircraft responds to behaviours exhibited by the flight crew. A crew culture that integrates concern for people and concern for performance contributes to safe flight.

The GRID®

Grid® theory provides a framework for understanding people through a two-dimensional grid describing concurrent and competing assumptions: concern for people and concern for performance. The interaction of the two concerns as illustrated in Fig. 7.2 yields different behavioural styles for individuals. The Grid depicts concern for people on the Y-axis ranging from 1 (low concern) to 9 (high concern). Concern for performance along the 'X-axis' also ranges from 1 to 9.

As you read the behavioural descriptions in Table 7.2 below, determine which style paragraph you think matches your own style. For additional clarification, ask people who work with you (or your family members and friends) to indicate which paragraph most closely matches your behaviour most of the time. The statements represent the Grid® elements of inquiry, initiative, advocacy, conflict solving, decision-making, and critique. Behavioural styles are indicated at the end of each description.

Developing the Future Aviation System

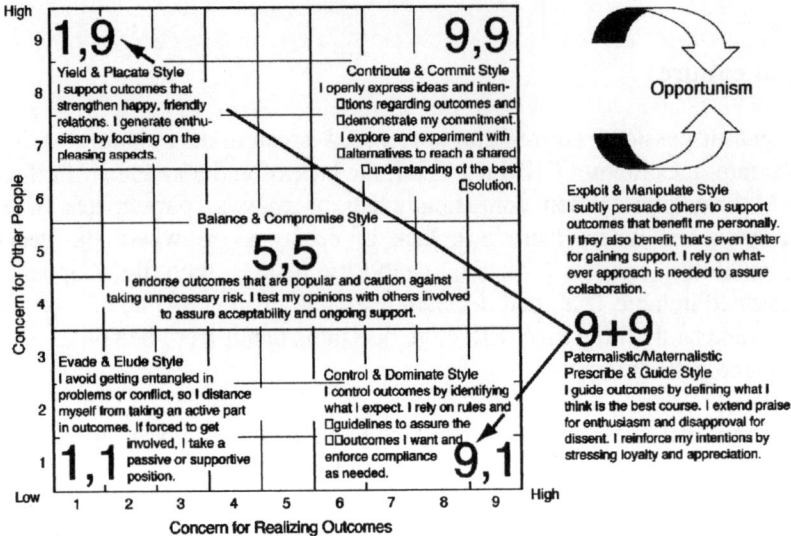

Figure 7.2 The Leadership Grid®
Copyright © Scientific Methods, 1998 from The Managerial Grid® by Drs. Robert R. Blake and Jane S. Mouton, 1968.

Table 7.2
Grid® Theory Behavioural Styles
Copyright® Scientific Methods, 1998 from Grid® theory developed by Drs. Robert R. Blake and Jane S. Mouton, 1968.

A. I let others make decisions or decisions are made by default. I am indifferent and usually go along with opinions, ideas, and actions presented by others, seldom expressing my own thoughts. When conflict arises, I try to remain neutral or stay out of it. I accept decisions, facts, and beliefs given me without much thought. When asked for information I pass it on. I avoid giving feedback even when asked (1,1).

B. To maintain good relations, I encourage others to make the decisions if possible. I embrace and support others' opinions, ideas, and actions even when I have reservations. I try to avoid generating conflict, but when it does appear, I try to soothe feelings and to keep team members together. I look for facts, decisions, and beliefs that suggest all is well. For the sake of harmony, I am not inclined to challenge others. I give encouragement. Praise is offered whenever something positive happens, but I go light on anything negative (1,9).

C. I search for workable, even though not perfect, decisions that others accept. I express opinions and ideas and take action in a tentative way so I can accommodate or avoid being too different. When conflict arises, I try to find a position that is acceptable. I am inclined to take things more or less at face value and only probe when an obvious discrepancy appears. I check out facts, decisions, and beliefs but not sufficiently to probe their objectivity in depth. I give informal or indirect feedback, which tends to be shallow (5,5).

D. I place high value on making my own decisions and am rarely influenced by others. I hold on to the opinions, ideas, and actions I think appropriate even though it means rejecting others' views. When conflict arises, I try to cut it off or to win my position. I investigate my own and others' facts, decisions, and beliefs in depth in order to be in control of any situation and to reassure myself that others are not making mistakes. I pinpoint any weaknesses or failure to measure up in order to place blame (9,1).

E. I double-check what others tell me and compliment them when I am able to verify their positions. I maintain strong convictions but permit others to express their ideas so that I can help them think more objectively. When conflict arises I terminate it and thank people for expressing their views. I give others feedback and expect them to accept it because it is for their own good. I have the final say and make a sincere effort to see that my decisions are accepted (9+9).

F. I place high value on arriving at sound decisions. I seek understanding and commitment. I feel an obligation to express my concerns and convictions; I listen for and seek out opinions, ideas, and actions of others; I am ready to change my mind when I am convinced. When conflict arises, I seek out reasons for it in order to resolve underlying causes. I actively collect and validate data, and continuously re-evaluate my own and others' beliefs, decisions, and facts for soundness and objectivity. I give feedback as appropriate during an operation. When the operation is completed, critique among crew members provides basis for diagnosis, review, and learning (9,9).

G. I dig out areas of vital private concern to me in an inquisitive but non-threatening way. I keep my own counsel but respond to questions when asked. I rarely reveal my convictions because then I don't have to stand behind them. When conflict arises, I shift and turn in an effort to circumvent it; I avoid getting caught in conflict head on. I use critique to motivate and inspire others to further action, which is in my best interest; I tend to discount negative aspects of performance as this lowers the level of enthusiasm. I lobby my point of view to 'sell' my position; I may use persuasion and indirect threat to ensure that my wishes are carried out (Opportunism).

Research indicates that people who complete this ranking activity will normally rank the paragraphs based on their *intended* behaviour rather than their *actual* behaviour. When people revisit the same activity after intensive personal effectiveness training, personal perceptions shift dramatically away from their original self-assessments. Understanding the gap between how a person intends to behave and how a person actually behaves is critical to understanding resistance to change.

A further illustration of this concept is in behavioural *alignment* toward safe flight. Research indicates that when flight crew members are presented with the preceding behavioural Grid® style statements (note Table 7.2), they select statement F more than 90 % of the time as the most desirable crew behaviour. If flight checklists are the culmination of best practices in workload management, then it could be said that best practices in effective crew behaviour promote alignment toward safe flight. If there is consensual agreement among people as to the effective use of Grid® style elements of inquiry, advocacy, conflict resolution, decision-making, and critique, the agreed-upon behaviours can provide a checklist.

CRM implementation

Pilots voice concerns as they complete CRM training. The principal concern is whether or not they can use the skills they have learned 'back on the job'. Having become aware of their own behaviours and what constitutes effective behaviour as studied and practised in simulations during CRM training, flight crew members are motivated to use their new skills. Data from Grid®s CRM seminar results show that prior to training 82 % of the individuals select the "9,9" style description ('F' of Table 2) as the leadership style they exhibit most often. After learning more about themselves through feedback and reflection on their own behaviours, only 25 % rank themselves as using this style most of the time.

The Gap

CRM training participants often comment that the organisation says one thing and does another. This does not come as a surprise based on research indicating a gap between what individuals think and say they are doing compared to what the consumers of their behaviour report. Each level of an organisational hierarchy or industry bears out the theory: there are gaps between perception and reality as individuals, teams, organisations, and industries. Culture is an extension of norms, which are the assumptions, attitudes, and values shared by a group of people. These norms drive behaviour so silently that people believe they are doing that which they are not doing and vice versa.

One CRM program history

United Airlines used Grid theory to design their CLR (Command/ Leadership/Resource) Management training in one of the first CRM training programs developed. Their intent was to stop human and hull losses. UAL worked to create a more effective flight deck culture through more participation. The captain retains final authority but with improved effectiveness through additional resources when required. All crew members feel free to not only participate in important decisions, but also, the obligation to do so. United Airlinesquote common language and alignment created by their CLR training contributed to safety during two well-publicised and dramatic stories: Flight 232 with its loss of all hydraulics and Flight 811 with a cargo bay door blown open in flight. Creation of a new culture enabled by more than 12,000 flight crew members trained in CLR contributes to safer flight for UAL. In fact, in the 17 years since the inception of CLR, United Airlines has not lost a single passenger due to flightcrew resource management error.

Integration of effort

When asked to rank their crew cultures, participants in a Grid® CRM seminar rank what they believe to be the most effective crew culture as well as their present culture. Normally this activity reveals a gap between the desired culture and the existing culture, between how things should be done and how they are actually done. If aviation crews know what constitutes an effective flight deck culture and management knows what constitutes an effective organisational culture, why is flight crew error the cause of 70% of worldwide commercial jet fleet accidents? Much of the answer resides in the interaction or resource management of teams, groups, and sub-systems involved. Integrated teams comprised of flight deck, cabin, dispatch and maintenance and purchasing, customer relations, FAA, JAA, ATC, aviation suppliers, and supporting organizations create generous opportunity for error in coordination of effort.

CRM and leadership

CRM principles used in crew training can also be applied from the top of an organisation down through the functional teams and expanded through the permeable boundaries of one organisation or system to another. Important to the effectiveness of culture change is the adherence to effective CRM principles and the subsequent development of an infrastructure supporting the desired behaviour. For example, if an organisation talks about the importance of effective teamwork and yet individuals are rewarded on an individual basis, the structure is not supportive of the new team behaviours. If an airline advertises safety yet builds an infrastructure that stresses speed and profit with a low regard for quality, the structure supports neither concern for people nor concern for performance. Safety is held captive.

Systems approach

Though the flight crew is named as the single largest contributor to worldwide commercial jet fleet accidents, this chapter advocates the flight crew as an important sub-system of overall transportation. Studying relationships within larger systems provides insight into what is required to create a culture of safety in aviation. Ongoing and effective team resource management among multiple and often competitive systems can begin to imbed improved performance through individual, team, and inter-team process. An integrated approach to organisation development is presented applying CRM skills not only to the flight deck, but also to the cabin, dispatch, maintenance, management, suppliers, customers, ATC, FAA, JAA, and other governmental systems. The aviation industry recognises the nested systems of global aviation transportation and has begun to look at the impacts created by the multiple interfaces. Integrated organisation development as depicted in Fig. 7.3 is iterative.

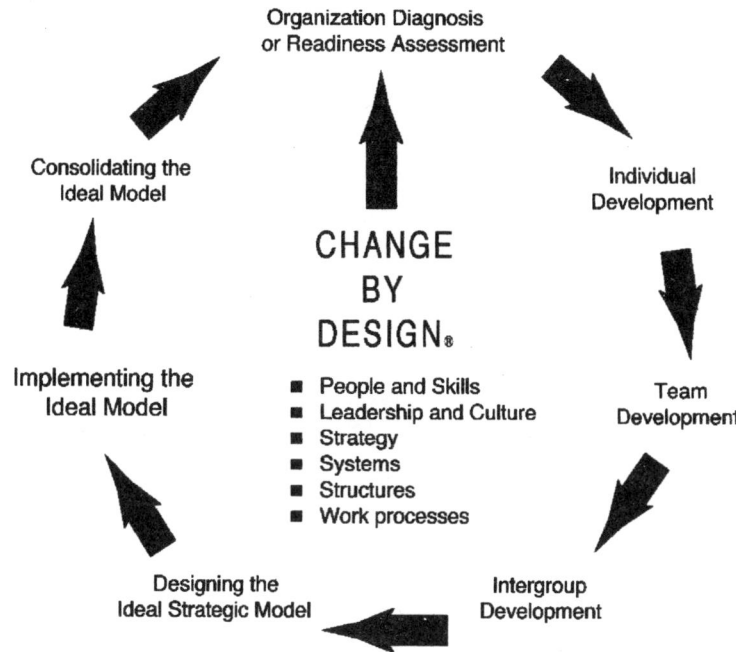

Figure 7.3 Change by design
Copyright © Scientific Methods, 1998 from Grid® theory developed by Drs. Robert R. Blake and Jane S. Mouton, 1968.

Future directions

When differing groups (aviation crews, management teams, and industry officials) interact, their competing interests evolve naturally. Miscommunication and ineffective behaviours can arise. These situations create an imperative for intergroup development among the aviation industry stakeholders. Formal intergroup development activities involving structured activities such as those provided in Grid organisation development are elegant in their simplicity. Each team identifies 1) the ideal model of working together, 2) how each team is currently working, and 3) ways to close the gap between actual and ideal.   Creating safe flight in the

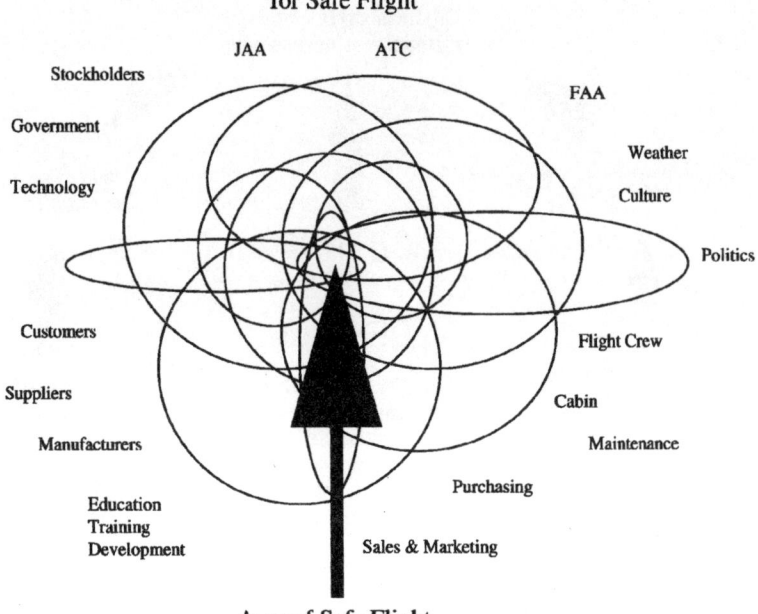

Figure 7.4 Aligned for safety or error?
Copyright Dianne Hill, Performance by Design, Austin, Texas, USA, 1998.

future requires the alignment of many discrete pieces constantly in motion. Alignment of the critical interfaces is the challenge represented in Fig. 7.4. Alignment for safety is critical.

References

Ashkenas, R., Ulrich, D., Jick, T., and Kerr, S. (1995), *The boundaryless organization: breaking the chains of organizational structure.* San Francisco: Jossey-Bass Publishers.

Blake, R. R., and Mouton, J. S. (1964), *The managerial grid*, Houston, TX: Gulf Publishing Co.

Blake, R.R., and Mouton, J. S. Command/Leadership/Resource Management Steering Committee and Working Groups, United Airlines

(1982), Cockpit resource management. Denver, CO/Austin, TX: *Cockpit Resource Management.*

Blake, R. R. and Mouton, J. S. Command/Leadership/Resource Management Steering Committee and Working Groups, United Airlines (1990), *Grid® cockpit resource management.* Austin, TX: Scientific Methods, Inc.

Blake, R. R. and Mouton, J. S. (1968), *Consultation: A handbook for individual and organization development,* Reading, MA: Addison-Wesley.

Blake, R. R., and Mouton, J. S. (1968), *Corporate excellence through grid organization development,* Houston: Gulf Publishing.

Carroll, J.E. and Taggart, W. R. (1987), Cockpit resource management: A tool for improved flight safety, In H. W. Orlady & H. C. Foushee (Eds.), *Proceedings of the NASA/MAC workshop on cockpit resource management (NASA Conference Publication 2455).* Moffett Field, CA: NASA-Ames Research Center.

Hackman, J. R. (1993), Teams, leaders, and organizations: New directions for crew-oriented flight training. In Wiener, E. L., Kanki, B. G., & Helmreich, R. L., (Eds.). *Cockpit resource management.* San Diego: Academic Press.

Hackman, J. Richard, Ed. (1990), *Groups that work (and those that don't).* San Francisco: Jossey-Bass Publishing.

Helmreich, R. L., Wiener, E. L. and Kanki, B. G. (1993), The future of crew resource management in the cockpit and elsewhere. In Wiener, E. L., Kanki, B. G., & Helmreich, R. L., (Eds.). *Cockpit resource management,* San Diego: Academic Press.

Hill, D. (1997), PIER Learning: People as Informal, Extended Resources. In: R. S. Jensen (Ed) *Proceedings of the Ninth International Symposium on Aviation Psychology.* Columbus, Ohio: Ohio State University, Department of Aviation.

Hofstede, G. (1997), *Cultures and organizations: software of the mind.* New York: McGraw-Hill.

Manningham, D. (1997), *'Does CRM Work?',* Business & Commercial Aviation December 1997, pp. 66-72.

Mouton, J. S. and Blake, R. R. (1984), *Synergogy: a new strategy for education, training, and development.* San Francisco: Jossey-Bass.

Statistical Summary of Commercial Jet Aircraft Accidents, Worldwide Operations 1959-1993 (1996), *Airplane Safety Engineering (B-210B)* Boeing Commercial Airplane Group. Seattle, Washington: Boeing, June 1997.

Yamamori, H. (1987), Optimum culture in the cockpit. In H. W. Orlady & H. C. Foushee (Eds.), *Proceedings of the NASA/MAC workshop on* cockpit resource management (NASA Conference Publication 2455). Moffett Field, CA: NASA-Ames Research Center.

The author

Dianne Hill, BA and MBA, is President of Performance by Design, an organizational development company experienced in culture change focused on developing collaborative relationships among independent teams within the dynamic global climate.

She conducts Crew Resource Management seminars based on GRID® theory using the elements of inquiry, advocacy, conflict resolution, decision making and critique. Locations have included the United States, Italy and El Salvador

Acknowledgements

I wish to acknowledge the generous permission of Scientific Methods, Inc. to use copyrighted Grid® theory graphics and tables illustrating key concepts and specifically the contributions of key staff members Karen McCormick and Rachel McKee. I thank Margaret T. (Peg) Shaffer of Paradigm, Inc. for introducing me to editor Rod Baldwin. My sincere appreciation goes to Capt. Stan Meader, UAL Retired, and H. Walter Barclay of SR International for their review and critique. I thank also additional reviewers including Sarah Barber, Eddie Jones, and J. Bruce Huffman who critiqued the initial draft. Dr. Harry H. Holzner of the DFS Air Navigation Services Academy in Langen/Germany contributed useful ideas through his advocacy of Grid® CRM principles. And, finally, I am deeply indebted to my mentor Capt. Tom Hudgens, UAL, Retired, for sharing his extensive aviation knowledge and promoting aviation safety not only through his many years of CRM seminar management, but also through his adherence to and advocacy of CRM principles.

8 Human factors in Air Traffic Control

V. David Hopkin

The inevitability of change

Many current air traffic control systems are functioning at or near their maximum traffic handling capacity, despite successful attempts to increase their capacity beyond what was originally planned. Although there is not complete agreement on how much air traffic there will be in the future, every prediction foresees major increases in air traffic, which current systems retaining present practices and procedures would be unable to accommodate. Therefore, to leave the present generation of air traffic control systems unchanged is not a practical option, even with their good safety record. Future systems must be able to handle more air traffic, preferably with enhanced safety and efficiency.

One apparent option can be discarded straightaway. To recruit more controllers and continue to partition further the region of airspace for which each controller or team of controllers is responsible becomes a self-defeating exercise because of the extra communications and co-ordination involved. Reduced to its essentials, the core human factors problem is to ensure that each controller spends less time in dealing with each aircraft, while maintaining, and preferably surpassing, the current high standards of air traffic control service provided. This objective must be achieved in a context where many other changes are taking place. Although these changes should benefit air traffic control, many of them introduce new human factors problems. These changes are of three main kinds which are considered in turn: changes in technology, changes in air traffic control concepts, and changes within human factors.

Changes in technology

Most technological innovations have occurred independently of air traffic control and not in direct response to its needs. They, therefore, have to be adapted to meet air traffic control requirements. The advent of information derived from satellites transforms the quality of the information about aircraft positions available to air traffic control. This in turn means that the absolute and relative positions of aircraft will be known far more accurately in the future, and any deviations from intended track should be detectable earlier. Also, this accuracy of information will become available in regions of the earth hitherto beyond radar coverage, notably over oceans. Separations between aircraft can then safely be reduced drastically, with the benefits of much increased transoceanic traffic handling capacity, improved fuel efficiency and shorter durations of transoceanic flights. The consequent changes in the controllers' tasks raise human factors issues, such as the depiction and usage of plan views of the traffic which may look like radar displays but are not derived from radar.

Currently, air traffic control still relies heavily on the exchange of information between pilots and controllers through spoken dialogues, the content and format of which are internationally standardised. Machines that can recognise human speech may have a future role in air traffic control, though the first applications seem likely to be confined to training. Spoken messages employ the ICAO alphabet internationally, but its words, devised to minimise phonetic confusion in humans despite noisy channels and mispronunciations, are not necessarily optimum also for machine recognition. Information other than message content is conveyed through human speech, and controllers form judgements about the speaker's competence, understanding and confidence from the pace, fluency, pauses, choice of language, accent, and responsiveness of the speech. This kind of information is lost when the data are transponded automatically.

Automated data links will transpond information between aircraft and the air traffic control system very frequently and accurately, reducing the need for speech. Data links can provide high quality information for computations within the ground based system or on board the aircraft, from which the safety of current traffic configurations may be checked and the safety of future configurations predicted. A human factors issue is how much of this transponded information must be known by the pilots and controllers. There is far too much to show them all of it, and a judicious condensation of it must be compiled for air traffic control purposes.

A notable future trend, also in response to ICAO policies, will be to consider communications, navigation and surveillance together. Combining technological developments allows this to be done. It has major implications for amalgamating human tasks, which have tended to be split under these headings in the past.

Technology has transformed the physical environments of some air traffic control workspaces, notably by providing displays which can be viewed in high ambient lighting. Much larger cathode ray tubes can be provided, with almost unlimited colour coding. Tabular information can be presented as windows in these large displays, with enhanced facilities for manipulating the windows and selecting their contents. Electronic versions of flight progress strips are becoming more common, a flight progress strip being formerly a piece of paper on which details of each flight were updated by hand by the controller following spoken exchanges of information with pilots or other controllers. It has proved difficult to replicate electronically all the functions fulfilled by manual flight progress strips, but it may not be necessary to do so. In human factors terms, it is not the same cognitively to update a flight strip by hand and to have it updated automatically. This raises the human factors issue of what the controller needs to know about automatic changes that are taking place, in order to match them and respond appropriately to them.

Input devices may also change. Touch sensitive surfaces may replace keyboards, and a mouse may replace a joystick or rolling ball. New labelling of functions may be needed, and there will be new relationships between displays and the input devices used to select, amend and update them. The practice of performing separate human factors evaluations of displays and input devices is yielding to the design and evalution of the human-machine interface as a whole. The merits of particular display types or codings, or of particular input devices, are then not an absolute property of them regardless of what they are used for, but their merits depend on the efficacy of the combination of displays and input devices within the human-machine interface in performing the designated tasks with them.

Technology can provide almost unlimited data storage. These data are used not only to reduce the controller's role in routine information gathering and data transmission but also to compile and compute aids and reminders to support the controller's cognition. Aids to cognition have to be introduced with some circumspection as they may be construed as rendering controller skills redundant or as whittling away human responsibilities. Hence, they may bring new kinds of human factors problems. Air traffic control, with or without cognitive aids, is complex, and it can be difficult to disentangle its cognitive aspects from its non-cognitive ones (Eurocontrol, 1996).

Changes in air traffic control concepts

The division of air traffic control responsibility between the cockpit and the ground based system is evolving. As a consequence, the interactions between the pilot and the controller must also change. Current and future forms of computer assistance have two main kinds of effect on the controller. One effect renders the controller's tasks more strategic and less tactical, with more emphasis on flows of traffic and less on individual aircraft. The other effect concerns prediction aids which emphasise what the traffic scenario will be in the future rather than what it is now. Together these trends promote the role of the air traffic controller as the manager of the traffic as a whole, with fewer direct interventions in the progress of single flights. Where possible, it is ascertained before the aircraft leaves the ground that its flight will not conflict with any other aircraft, and, if it will, its departure time is amended to remove the potential conflict at its source.

This attempt to make air traffic control a seamless process for each flight from take-off to landing blurs some traditional divisions of responsibility in air traffic control, particularly regarding the handover of control responsibility between en route, terminal area and approach controllers during the flight. A tactical intervention by a controller in the earlier stages of a flight may not merely resolve a particular problem but could invalidate the planned conflict-free further progress of the flight. Controllers therefore need to be able to discover the possible longer term consequences of their intended actions.

A comparatively recent extension of this thinking is the concept of free flight. Within its short history, this concept has already evolved considerably. Its essential intention is to allow the pilot to be freed from the traditional constraints imposed by air traffic control such as its route structure. The pilot should be able to request a particular departure time and a flight profile following the most direct route (normally a segment of a great circle), in the expectation that air traffic control would normally grant this request, having checked that no conflicts would accrue from it. Some free flight proposals place on the pilot the responsibility to reach a designated position near the arrival airport at a designated time, as a condition for being granted clearance to proceed immediately to land with no delays or diversions. The pilot has the responsibility to choose a departure time, speeds and flight profile to achieve this objective under the prevailing weather and other conditions.

Where routes are retained, a series of slots along a route are allocated to aircraft flights to ensure safe separations within the route and to permit traffic planning in accordance with route capacities for handling traffic. Slots are also a tool for the advance planning and clearance of flights. Developments in air traffic control such as slots and free flight have major effects on the future roles of air traffic controllers and hence on the tasks which they will have, many of which are still being worked out.

Changes within human factors

The applications of human factors as a discipline to aviation in general, and to air traffic control in particular, have expanded significantly in recent years and are continuing to do so. The history of human factors studies on air traffic control covers nearly fifty years (Hopkin, 1970), but the human factors resources devoted to air traffic control have always been much smaller than those applied to cockpits. Some current applications, for example to the verification and validation of the functioning of air traffic control systems, to certification, to maintenance, and to cultural differences are all quite new, and the emphasis on treating air traffic control as a team activity in order to resolve some of its human factors problems, is also comparatively recent.

While air traffic control remains substantially manual, controllers develop skills in planning and organising the traffic tactically themselves. In its manual forms, air traffic control is highly observable. Although the activities within the air traffic control workspace may be well-nigh meaningless to an onlooker with no knowledge of air traffic control, a professional colleague or supervisor can understand very well what is happening by watching what the controller does and listening to what the controller says, to the extent that several other important functions rely on this observability. For example, it is used to assess the controller's professional competence, to establish each controller's professional reputation, to judge when a controller needs some help, and to gain the respect and trust of colleagues as a fully accepted member of the control team.

An incidental consequence of many forms of automation and computer assistance in air traffic control is to render the controller's roles and activities much less observable, to the extent that human incompetence or inadequacy may be concealed successfully from others. Traditional forms of supervision that depend heavily on observation may become impractical. A human factors problem is to establish whether it is necessary to compensate for these unintended effects of automation, and to devise and test effective means to do so if they are needed.

Technological innovations expand the range of human-machine relationships that are possible. The option of entirely manual functions remains, though it may become rarely used. A hierarchy of relationships develops in which the human roles and initiatives become progressively more marginal until they may vanish altogether. This changes the kinds of human error that are possible and the human responsibilities and interventions that are feasible. An incipient human factors problem is that the controller may no longer retain full control over functions for which he or she remains legally responsible.

When computer aids can select, extrapolate, compute, collate or predict information far more quickly and accurately than the unaided controller can (and indeed this is their main justification), it follows that the controller cannot check their correctness in detail in the time available but must learn to trust them. It also follows that the controller may be unable to act as an effective back-up to them if they fail because they are so much quicker and more accurate than the controller can be. Although the controller is expected to support machine objectives, this is often a one-way form of support as most machines cannot lend comparable support to human intentionality and the attainment of human objectives. Usually, no provision is made for the intentions of the human controller to be conveyed to the machine in order to enlist its help in achieving them, but this may represent a longer term advance in air traffic control. Such problems, and their legal implications, are live human factors issues.

Some human factors issues in air traffic control have always had to be treated internationally. The same aircraft and pilots flying to many countries need a core of standardised practices and common information exchanges. Air traffic control systems and equipment are made by a limited number of manufacturers but marketed throughout the world. The tasks and functions within them imply certain command structures, divisions of responsibility, and training schedules to instill appropriate prior knowledge of the system and its functioning among the controllers who will use it. These divisions of knowledge, work and responsibility may not be compatible with the practices of other cultures in regard to work sharing, delegation of responsibility, and taking initiatives. This may make them difficult to teach, and at odds with aspects of the legal systems of other countries. Such issues, commonly called cultural ergonomics, have already arisen in relation to cockpits and are expected to become more important in air traffic control.

Human factors has become less fragmented and more unified as a discipline. It is now more widely recognised that the solution of any human factors problem will have further human factors implications within the system, and that these must be recognised in advance lest one solution merely generates further problems elsewhere. One example concerns selection and training, treated separately in the past but now seen as linked. Can deficiencies in the selection process be countered by modifications to training or will the quality of the training product inevitably be lowered if the selection process picked the wrong people? More broadly, are both the selection and training processes for controllers pre-judged by the success or failure of the recruitment procedures to attract the best applicants for jobs in air traffic control, and could the combined efforts of recruitment, selection and training be undermined by the allocation of newly qualified controllers to inappropriate jobs or by subsequent inept handling of the controller's career development? Such human factors problems are often made more complex because they cut across traditional divisions of administrative responsibility.

Developments in human factors

Human factors as a discipline has gradually gained wider recognition as having an essential contribution to make to the safety and efficiency of large human-machine systems. Within aviation its range of applications has expanded to include, for example, security procedures, aircraft evacuations in emergencies, and the jobs of aircraft cabin staff. Within air traffic control its range of applications has expanded to include, for example, the development of attitudes to automation, the consequences of different shift rotations, changes in human understanding and memory as a result of automated aids, and the mainsprings of esteem and pride in the profession. Human factors in air traffic control has become accepted as reputable in its own right, with a recent series of texts on it (Hopkin, 1995; Cardosi and Murphy (Eds), 1995; N. R. C., 1997), with its representation in a series of international digests (ICAO, 1993), with its inclusion in texts on aviation human factors (Hopkin, 1988) and in texts on human aspects of automation in large systems (Hopkin, 1997), and with coverage of it in plans for aviation human factors research (F.A.A., 1991). The burden of proof has changed from gaining acceptance for human factors to delivering all that it promises. The latter is the challenge in future aviation systems.

A corollary is that human factors studies in air traffic control cannot afford to be an isolated activity. Comparable problems are likely to have arisen, and may have been solved, in other large human-machine systems or in other aviation contexts. Foreknowledge of these problems and of their solutions alerts the human factors specialist on what to look out for in air traffic control, and requires professional judgement on the degree of commonality between air traffic control and other systems and on the probability that findings obtained elsewhere will remain valid if extrapolated to air traffic control. The aim is not just to save effort and avoid duplication. Competent professional human factors resources to address problems in air traffic control are in short supply and they are never sufficient to perform all the human factors work that should be done. Therefore any guidelines on how to deploy these limited resources most productively should be welcomed.

Currently many nations seem to lack a policy for the overall allocation of the limited human factors resources available for air traffic control. For example, those who work on selection, on training, on workspace design, on performance measurement, and on team roles and functions may all be seeking funding independently of each other, without any apparent overriding policy on the proportion of human factors resources that should be deployed on these various applications to obtain best value for money. The relative priority accorded to different kinds of human factors problem in the past has over emphasised problems that are easy to recognise and to tackle and under emphasised those that are less tangible and definable but may nevertheless have more influence on the safety and efficiency of the air traffic control system. Human factors contributions have often been made too late in the system evolution when decisions that have already been taken preclude optimum human factors solutions. Future developments are expected to include more consideration of the optimum deployment of total human factors resources, to introduce human factors contributions from the earliest stages of the air traffic control system evolution, and to make further human factors contributions iteratively as the system becomes progressively more defined in detail.

Expansions of human factors roles have been accompanied by additional human factors measurements. Recognition of the relevance of some of them has been comparatively recent but will probably prove enduring. The traditional purpose of air traffic control has been the provision of the safe, orderly and expeditious flow of air traffic, and traditional measures of system functioning and controller task performance have therefore been derived from this purpose. These cover system events and controller actions, and their timing, sequencing and correctness. Some of these are now interpreted in terms of human-machine compatibility. Human errors, their causation by the system, and their effects on the system, have always been of great importance because of their implications for safety, and various taxonomies of human error in air traffic control systems have been devised (e.g. Stager, 1991).

General taxonomies of human error usually refer to aspects of human thinking and information processing, and they may have to be adapted to make them more compatible with the kinds of emergencies, non-standard situations, and existing categorisations of incidents which pose a potential or real hazard in air traffic control. These existing categorisations have been derived from other frames of reference, and they may not readily match those based on human attributes, but it is essential to match them sufficiently to permit cross-referral. A useful development has been confidential human factors incident reporting whereby the occurrence or potential occurrence of a hazardous situation can be reported in confidence and is disidentified before any dissemination, without any blame or career or management implications for the person reporting. People can report in this way their own errors, and errors which the system would not have prevented or recognised. This can aid safety by tapping into sources of human error that may not emerge in any other way.

Air traffic control has a series of further objectives that are implicit rather than explicit. Some of these, like fuel conservation and noise abatement, are not human factors objectives, but many are, and these include low labour turnover rates, minimising training costs, job satisfaction, the design of effective team roles including those of supervisors and assistants, and means to exercise legal responsibilities. When forms of computer assistance progressively evolve towards full automation of human functions, the human controller becomes a planner and manager of resources rather than an intervener in tactical aspects of air traffic control. It becomes less feasible to retain enough human involvement in tactical processes to keep the human's understanding of the full scenario up-to-date, so that the practicality of employing the human as an effective back-up in the event of system failure is undermined. Attempts to intervene can only be made through the human-machine interface which may make no provision for unorthodox actions by the controller attempting to resolve emergencies, or the controller's unusual actions may be overruled by the machine which treats them as invalid. The balance of responsibility between human and machine can often be gauged by examining the relative extent to which the human controller or the machine can introduce initiatives.

Much of the data in standard human factors handbooks (e.g. Salvendy, 1987; Boff and Lincoln, 1988) can be applied in air traffic control, but usually the recommendations require some modifications which must be done by a skilled human factors specialist with good knowledge of air traffic control. Among the commonest reasons for modifications are controller's eyesight standards, ambient lighting, flexibility of staffing levels in workspaces, shared displays or input devices, methods of watch handovers, and special features of air traffic control tasks which have no counterpart in other large human-machine systems. Most of these reasons will remain as applicable in future systems as they are in current ones.

As more computer aids for the controller's higher mental functions are provided, more of the characteristics hitherto treated as uniquely human become machine attributes also. Examples include intelligence, adaptability, flexibility, and a capacity to innovate. Principles are still somewhat rudimentary for the optimum matching of the human and the machine when both possess these attributes. When these principles have been clarified and validated, they would seem to entail revisions in both the selection and training procedures for controllers. The problem of teachability has already arisen with some of the more complex forms of assistance, if their functioning is to be understood well enough by the controller for them to be used as intended. Those who devise new computer aids may be the only ones ever to utilise their full potential because too many aspects of them are not user friendly or self-evident enough to be readily teachable.

The literature on many other aspects of work and on working conditions refers primarily to jobs which are either simple and highly repetitive or which rely heavily on human skills and direct interventions with feedback of results. What happens to loyalty, self esteem, morale, interest in the work, and identification with the job, profession or place of work, if many of the existing controller skills are rendered redundant? Most controllers enjoy air traffic control in its more manual forms and find it satisfying. At smaller airfields it will not pay to introduce sophisticated technology, and so there will always be a need to integrate some aircraft under more manual forms of air traffic control into these more automated systems. These are further examples of the enlarged range of human factors problems in air traffic control which will have to be tackled in the future because of the introduction of more advanced forms of automation and computer assistance. For the reasons stated at the outset, these problems will not go away and cannot be allowed to drift because air traffic control cannot remain as it is but must evolve.

Normally, major changes in air traffic control, such as those that have taken place or are pending, would be expected to change the selection procedures for controllers at some stage by requiring new attributes in controllers and by rendering some existing attributes redundant. If new forms of computer assistance can build on existing controller skills this is an advantage but it is not always practical. Although a few tests have been devised specifically for air traffic control, many of those used in air traffic controller selection have not, but are measures of fundamental human attributes deduced to be relevant to air traffic control and applied to it retrospectively. This raises contentious issues of how relevant they actually are, and whether all air traffic control jobs are sufficiently similar to each other for the same tests to suffice in the selection procedures for all air traffic control jobs. Perhaps test batteries should differ for different air traffic control jobs, or a common test battery should be employed partly to allocate candidates for training in different jobs. These are currently live human factors issues in controller selection.

Despite, or because of, future automation, most predictions suggest that the number of air traffic controllers in the future will remain much as it is now. The responsibilities, roles, jobs, and tasks will change, but the traffic demands will rise. The increase in automation and technological advances will enable more traffic to be handled rather than permit fewer controllers to be employed. But each controller will therefore spend less time in dealing with each aircraft.

The human-machine interface

Much practical data in human factors handbooks (Sanders and McCormick, 1993) and in some air traffic control human factors texts (e.g. Cardosi and Murphy, 1995) concentrate on aspects of the human-machine interface and the workspace around it. Although traditionally the two major aspects of the human-machine interface, namely displays and input devices, have often been designed and evaluated separately, the current and future emphases relate more to the interface as a whole. There are several reasons for this. One is that the usage and efficacy of either a display or an input device depends on the relationships between them as well as on the characteristics of each. Another reason is that communication is an aspect of the human-machine interface which is integral to it and has to be considered along with the displays and input devices through which it is effected. A third reason is that some technical devices such as touch sensitive surfaces are both diplays and input devices and it is impractical to separate these two aspects of them when they are being evaluated. A further reason is the increase in human factors studies of the more strategic aspects of air traffic control and on what it is achieving, which places less emphasis on the detailed means by which specific actions are achieved through using detailed display contents and specific inputs. There is often some lag in incorporating these new ways of treating human-machine interfaces into handbooks covering their design and applications. But the new roles and tasks envisaged for air traffic control imply significant modifications of existing human-machine interfaces, and in some cases new interfaces altogether.

The displays within the human-machine interface are the main means by which information is conveyed from the machine to the human controller, and the input devices within the human-machine interface are the main means by which the controller conveys information to the machine and enters it into the system. In the past, designers of the human-machine interfaces in air traffic control have not sought to make the information on the displays or the operation of the input devices self-explanatory. Air traffic control interfaces generally contain very little information about what they are, what they are for, what are the functions and tasks to be fulfilled with them, what human errors in operation might be possible, or what forms of feedback are provided for the user. Controllers have to learn the meaning of all the display contents, the functions of all the input devices, the tasks to be performed, and when and how to execute them. The system works because of what the controllers know. In current and planned future systems, there would be no air traffic control in the absence of controllers.

This raises the issue of whether a future use of computers in air traffic control should be to make the processes and procedures of control more self-evident. As usual, there are arguments in favour and against this, and the decisions depend on where the balance of advantage seems to lie. Some kinds of aid inevitably have the effect of making actions more self-evident though that is not their prime intention. For example, wherever a menu is presented for the controller to make a choice, this introduces constraints about which action should be done next and the choice of actions available. More self-evident actions should reduce training time and training costs. However, if too much is made self-evident so that the machine leads the controller through sequences of actions, this can be experienced as deskilling, loss of flexibility, or loss of responsibility, all of which would encounter problems of acceptability among controllers. In principle, many of the human-machine interfaces within air traffic control could be redesigned to function more as teaching machines, and the desirability of such a change is debated.

The specification of computer aids for the controller can often include the presentation of options, whereby the final decision, and hence the responsibility, is left with the controller. What are not normally provided are the reasons for the choice of presented options. The controller may either have to take these on trust, or use the human-machine interface to enter into dialogues with the machine to try to discover what the reasons are, in order to decide whether they are valid in the circumstances or whether all the presented options should be rejected because of some information or instructions known to the controller but not to the machine.

Another possible function of automation is to use the displays to query a human action or to inform the controller that a particular human action will not be executed because the machine treats it as invalid. To query a human action has become common wherever it could have major consequences such as the deletion of large amounts of data, but there is usually provision to override the query easily in such circumstances. The actions that a controller can perform in an emergency will be governed by the facilities within the human-machine interface that are relevant to dealing with it and by whether the controller remembers how to use them, or is reminded by the machine how to use them. Current and proposed forms of computer assistance and interface design are not very good at conveying human intentions to the machine. Any provision of automated means to veto human actions poses the dilemma of whether the controller should have a facility to override such a veto in emergencies.

The human-machine interface contains relationships between the displays and the input devices. Standard human factors handbook recommendations can usually be adapted to make these relationships obvious. High reliability is essential, as feedback is normally provided through these relationships. Relationships between displays and input devices can be shown by several means. Among the commonest are their respective locations within the workspace, common layouts for displays and input devices, common coding of related information on displays and input devices, uniform labelling, immediate and not delayed feedback between actions and their effects, highlighted information on both displays and input devices, and the duplication or linking of options on displays and input devices.

Display and input device relationships can also demonstrate through layouts and codings how functions are related, and how tasks are grouped together. Both displays and input devices can show sequences of tasks and sub-tasks, and suggest task priorities. They can help to indicate which tasks are included in the job done at a particular work position, and how the tasks are grouped together. They can also help to reveal divisions between tasks, between functions and between responsibilities within the workspace of an air traffic control suite. Decisions about the relationships between the displays and input devices within a suite, combined with decisions about the specification of the human-machine interfaces, are crucial in determining the extent and manner in which the work may be shared among teams, and the practicality of direct supervision of the work.

The design of the human-machine interface should permit a smooth transition between associated tasks, and give some indication of the relationships between them. In systems without much automation it is usually possible to make an informed judgement on how burdensome the current and pending task demands are likely to be on the controller. Some information to aid such judgements usually remains in more automated systems, though less than before. Wherever task changes are introduced, corresponding changes in the controller's workload are almost inevitable, and the means of exercising control over that workload, and the extent to which such control can be exercised at all by the controller, are usually changed. It becomes necessary to check that new forms of computer assistance do not add unduly to controller workload through the task changes that they introduce.

Task changes in the form of automated aids often also alter the causes of stress among air traffic controllers, particularly if they force the controller to use aids that are not fully understood or not completely trusted, and this too must be checked when changes are proposed. The means through which the controller's legal responsibilities are exercised must not be inadvertently whittled away by denying the controller access to pertinent information that was formerly available or by preventing some human actions and initiatives that formerly were possible. If aids which have such effects are introduced, corresponding changes need to be made in assigning the legal responsibility for what occurs. There is an incipient danger that the controller may retain legal responsibility for activities which the controller no longer has any power over. The reduced role of speech may be implicated in this.

Although it is not always acknowledged or even recognised at the time, design decisions about aspects of the human-machine interface are also decisions about the kinds of human error which are possible and will ultimately occur in its use. The errors that are feasible in making even the simplest kinds of data input are not the same with a keyboard, a mouse or alternative input devices. Similarly, the errors that occur in reading the alphanumerics on displays depend on the character size, case, layouts and font, on their usage as determined by the tasks, and on any potential similarities between what is expected and what is presented. Many human errors could be prevented by better recognition that decisions about error sources are inherent in many design decisions.

A further way of controlling the incidence of human errors introduced when systems are updated is to examine the degree of correspondence between the new system and the old one in terms of their human-machine interfaces. Controllers will have become familiar with the old interface and as much direct transfer of learning as possible to the new interface should be encouraged. This helps to reduce human input errors, and also reduces training. Under stress or high task demands in particular, there is a tendency to revert to old habits, and errors result if there is a change in the function of keys at familiar positions. It is better to re-cast a keyboard completely than to retain many keys in similar positions but change their functionality. Extensive work has been conducted on how to train controllers in the use of new equipment, but very little work has been done on the associated problem of how to train people to forget long-familiar procedures and skills which are no longer relevant. A practical aim is to try to ensure that if these old habits are not forgotten, their unwanted recurrence would at least be detectable and not dangerous.

The tendency in air traffic control is to replace a radar display of modest size and other associated displays of tabular and ancillary information with fewer displays but including a much larger radar display within which are windows showing tabular information. A further tendency is to replace paper flight progress strips annotated by hand with electronic versions of them updated through keyboards, while retaining much of their original functionality. Both these tendencies require input devices which permit the manipulation and amendment of data.

Automated conflict detection, whereby aircraft are highlighted if their predicted future paths will infringe permitted minimum separation standards between aircraft, are becoming more common, and automated conflict resolution will become more common in future. The latter provides one or more computer solutions to the predicted conflict. Initially the controller can choose which to adopt and the system will check and may accept an alternative solution devised by the controller, but difficulties of responsibility arise if the predicted conflict position and the present aircraft positions are all the current responsibility of different controllers. It is feasible in the longer term to consider conflict resolution as a fully automated process which the controller is not informed of and for which the controller therefore cannot be held responsible, which would change and reduce controller roles quite dramatically. Many such forms of computer assistance introduce new time frames for controller interventions, as well as changing the interventions that are possible.

Some human factors developments depend on technical advances which are relatively independent of computer aids. An example is the advent of colour coding on air traffic control displays. This became practical at a time when satisfactory monochrome codings had already had to be devised for all coding dimensions of high operational significance. Therefore colour coding was often competing with established monochrome codings which had already undergone progressive refinement to meet operational needs. At first, colours tended to be oversaturated and garish, and sometimes lacked sufficient brightness contrast with the display background to be read easily and comfortably. They were nevertheless generally liked by controllers, even when it could be proved that they conveyed no benefits in terms of task performance. Further technical advances have extended the applications of colour in air traffic control, so that more recent examples have utilised display developments to employ colour in more subtle and sophisticated ways that take account of its unique coding advantages. A recent example is the application of transparency principles to the portrayal of air traffic control information on radar displays with superimposed alphanumeric and symbolic information (Reynolds, 1994; Hopkin, 1994a).

Air traffic control at main airports, over oceans, and on main air routes has to contend with gross but generally predictable changes in demand, associated for example with time of day. It also has to be adaptable to sudden changes in demand, both at a detailed level such as when a runway at an airport has to be closed at very short notice, and at a broader level, as when a heavy snowfall or severe storms close many airports across an extensive region at about the same time. There has to be some capability to split or amalgamate air traffic control jobs, sometimes more than once, to cater for gross changes in demand, so that the system is fully staffed to meet maximum demands but can be effectively staffed by small numbers of controllers when traffic is light, as it often is during the night. Human-machine interfaces therefore have to allow many jobs to be split or amalgamated, a complication which is common in air traffic control but comparatively rare in other large human-machine control systems. Associated requirements which the human-machine interfaces must meet are to allow efficient team activities, to permit some equipment sharing, to allow controllers to intervene to assist colleagues who are dealing with excessive demands or emergencies or are having other difficulties, and to give each controller some means to contact colleagues and ascertain what they are doing. Many control suites are designed so that the staffing level on the suite is itself flexible, which means that there must be room for several controllers at the suite under some circumstances whereas one controller can staff the suite without undue stretching and movement under other circumstances.

At one time air traffic control environments had to be quite dark because of the brightness limitations of unprocessed radar displays. Now, processed secondary radar displays and advanced technology allow the ambient lighting of the typical control room to be quite high, often at about the minimum level of lighting recommended for offices. Ironically some of the largest displays have temporarily re-introduced some restrictions on the ambient lighting because of their limited brightness capability, but this is not expected to be a long-lasting constraint. Several features of the profiles of air traffic control suites are commonly adjustable now, and workspaces favour a modular construction. As a consequence the suites and workspaces for different air traffic control jobs can appear superficially to be very similar, because alternative modules are of standardised sizes and shapes and many of the differences between jobs are represented by differences in software rather then hardware. Human-machine interfaces with different functions within the same system have therefore tended to become more alike in appearance.

If ancillary and rarely used displays, for example of cartographic information, are placed on top of the radar displays, the suites may become so high that the seated controllers cannot see general wall-mounted displays over them, and have some difficulty in seeing which other positions are actually staffed and at what level. Some lines of suites are curved, which can help controllers to view wall-mounted displays more directly, at the cost of restricting how much each can see the activities of colleagues further along the suite. In such circumstances the human-machine interface may have to include facilities through which the controller can discover staffing levels and activities elsewhere, and can contact colleagues to co-ordinate communications or transfer calls within the system or beyond it.

In more manual air traffic control systems, the supervisor can exercise direct supervision, or even interventionist supervision, because the system specification makes provision for this. As automation is introduced, more of the controller's activities are through the human-machine interface and fewer with colleagues. Some traditional forms of participatory supervision become impractical. The role of the supervisor is not always defined concurrently with the roles of the controllers, and may be changed as a consequence of the controller's new roles rather than be formulated from the outset in conjunction with them. A problem is to ensure that real as distinct from nominal supervision is still possible, if that remains the guiding policy.

If the supervisor's role approximates to that of a watch manager when it has previously involved detailed supervision of controllers' activities, an extra safety back-up may have been inadvertently removed. The supervisor may have additional responsibilities yet have access only to the information and aids available to the controller, so that what takes place is a transfer of responsibility rather than effective supervision. For effective supervision, the supervisor generally needs access to facilities beyond those that the controllers have. This implies different or extra facilities at the supervisor's human-machine interface.

The workspace design for air traffic control, including the interfaces within it, must accommodate the other activities apart from air traffic control that are performed in that workspace. These can include teaching, training, assessment, demonstrations, handovers of responsibility, supervision, assistance, reliability checks, maintenance, and cleaning. Some of these, such as assessments and reliability checks, may require additions to the human-machine interface to make them possible. Others, such as demonstrations and cleaning, may be difficult to perform efficiency unless their needs have been foreseen and allowed for. The general effectiveness of the human-machine interface for its many purposes should also be examined as part of the certification of the whole system and of the interfaces within it.

Human factors and certification

Certification is a skilled human job. It therefore lies within the province of human factors in that the work it entails depends upon human capabilities and limitations (Wise and Hopkin, 1997). The application of human factors to certification was included in the United States National Plan for Aviation Human Factors (Federal Aviation Administration, 1991), and others have advocated it from time to time, but a recent report on human factors in air traffic control (National Research Council, 1997) does not include references to research on human factors aspects of certification in air traffic control for there is little to refer to, and it does not discuss certification extensively although it does cover the topic. In fact, there has been little progress since the National Plan appeared, except for a conference on human factors certification, the proceedings of which have been published (Wise et al., 1994). This aimed to make a start by bringing to light what relevant work had been done, by defining if and how human factors could be applied to certification, and by gauging what new human factors problems might be expected to arise and what familiar ones would recur.

Some of the origins of this conference on certification can be traced to two earlier conferences, one on human factors aspects of automation and systems issues in air traffic control (Wise et al., 1991), and the other on human factors issues in the verification and validation of complex systems (Wise et al., 1993). The diversity of practices and procedures that had arisen among the various disciplines with a legitimate contribution to make to the design, evolution and evaluation of complex human-machine systems meant that there were no practices and procedures which could command universal acceptance for the purposes of verification and validation and none that could deal with every aspect of the fully functioning system, including the many significant interactions within it. A trap for the unwary was to use the same data both to formulate and to test hypotheses. Processes such as verification and validation seemed to imply measures and criteria that were not only independent of the system itself but must be comparable to the system in complexity.

Various validation procedures had evolved for different aspects of the system, such as testing the reliability of components, establishing the predictive value of the selection procedures for controllers, de-bugging software, and many others. Each in their way did make useful practical claims and assessments of the validity of the functions to which they were applied, but none could be applied to the others or to the system as a whole, and they were never intended for usage beyond their original applications. Most were applied at an intermediate stage, after the system had been designed but before it became fully operational, and they varied greatly in the extent to which they could be changed or scrapped if their validity proved to be low. Generally, they were empirical and lacked a theoretical rationale, so that the validity sought was what seemed reasonable and was not usually determined by absolute criteria that must be attained. Often there were no independent criteria which could be used either to indicate where further improvements in validity were possible and must be sought, or to indicate where further attempts to improve validity would probably be fruitless and should be abandoned.

One approach was to treat validation as an aspect of system design, and some implications of this were pursued. An alternative was to examine certification as a form of validation which seemed to meet some of its requirements, notably independence from design, applicability to the functioning system as a whole, and the unification of different practices. It also has the merit of being a legally sanctioned procedure. Issues on which concern was expressed include its degree of generalisability, and the potential of bias towards whatever was included in certification and away from what was not and might therefore be comparatively neglected with impunity. When certification is practised, it is normally applied to a new system about to become operational, to a system that has had major revisions, or to a system about to be re-started after being unserviceable.

The relationship between human factors and certification can be of two distinct kinds. One approach, which would not be expected to lead to radical changes, is to apply human factors to current certification procedures as they stand and to optimise them as procedures in human factors terms, without questioning whether they are the best procedures for achieving human factors certification objectives. The other approach, which would be much more likely to recommend radical changes is to scrutinise and evaluate current certification procedures in human factors terms, considering them as human work and attempting to match them optimally with known human capabilities and limitations. It must be emphasised that this second approach is not a human factors bid to take over certification or to re-cast it, just as the application of human factors to air traffic control is not a bid to replace controllers with human factors specialists. However, applying human factors to certification as a human work activity can be expected to bring benefits to certification comparable to those that have accrued when human factors have been applied to such contexts as aircraft cockpits and air traffic control systems. The application of human factors is not merely the use of the data in human factors handbooks, though that is a part of it, but is also the introduction of many relevant human factors concepts and frames of reference such as human-centred automation, team functionality, situational awareness, taxonomies of human error, and relevant cognitive concepts (Hopkin, 1995).

One human factors issue in relation to certification concerns how to certify the certifiers, which is partly a matter of training and partly depends on the tight definition of the procedures to be followed. Clearly the outcome of certification should not depend on which particular certifier carries it out, but the procedures should be sufficiently standardised and devoid of ambiguity that their outcome is independent of which qualified certifiers actually do the work. Similar human factors problems have arisen and have had to be solved in many other contexts, and the experience so gained should be helpful when applied to certification. This issue is a particularly complex one because currently certifiers are encouraged to rely on their own initiative and experience, and not to follow procedures that are rigid.

Another looming human factors issue in regard to certification concerns the nature and extent of the forms of automation and computer assistance that should be applied to the certification procedures themselves. They will not escape the current trend favouring more automation. Potentially the application of a greater degree of automation to them should facilitate standardisation of them, but the familiar human factors issues of the division of work between human and machine and the retention of human legal responsibility when some procedures are automated are bound to arise. Who or what to blame for any failures or deficiencies becomes a more recalcitrant problem as more automation is introduced into certification.

The introduction of human factors into certification implies that certification must be able to deal with human factors concepts such as tasks, skill, workload, stress, and the many effects of the system on those who work within it. Many aspects of the system are certified. These include not only the air traffic control system itself as a means of controlling air traffic, but various components, software, procedures, rules, and information sources. Also certified are the people with jobs in the system, that they are qualified to do their jobs, and the training that they have successfully completed in order to become qualified. For people, the certification may be full certification, or conditional certification that allows the controller to function only under the close supervision of colleagues who are fully certified.

Different procedures are followed for these various kinds of certification, and they do not fit neatly together. Nor do the relevant human factors concepts fit neatly the traditional divisions of certification, between equipment, procedures and personnel. However, the application of human factors to certification can benefit through the lessons learned from its many previous applications in other contexts, so that, for example it would examine the underlying mental and cognitive processes that are not directly observable and it would not be restricted to the more superficial manifestations of these underlying processes as represented by observable and recordable human activities (MacLeod & Taylor, 1994).

An objective in applying human factors to certification should be to exemplify all that is best in human factors practice (Hopkin, 1994b). This implies the need for skilled human factors judgement about the validity for certification of existing human factors databases, which are often based on simpler contexts. Not only must the human factors specialist be able to work productively in close harmony with those whose job it is to certify the system or various aspects of it. It may be of value for both human factors and certification to have a highly experienced and specially trained controller, the air traffic control equivalent of a test pilot, as the third member of a closely integrated team, with special training and skills (Westrum, 1994).

Certification can be treated from the point of view of its content, but also as a process or as a product. Although there may be conditions attached, ultimately it is a pass/fail decision: certification is granted or it is not. Certification may denote that a required level or standard has been achieved or that performance is within acceptable tolerances. Future trends are likely to include certification procedures that are more remote and less disruptive of system functioning, certification where the automation of processes is applied to progressively more complex aspects of system functioning, and a change of emphasis in system designs so that some certification can rely on built-in diagnostics that act as certification tools. The processes of verification and validation should help to build the user's trust in the capability and reliability of the system, and certification is one means of formalising that trust. If it engenders trust that proves to be misplaced, then this may reflect adversely not only on the certification processes themselves but also on those who carry them out and on the automated aids that they have used for certification. It takes a long time and evidence from many incidents to build trust, but a short time and evidence from very few incidents to destroy it.

Conclusions

Although air traffic control must evolve, changes in it are likely to seem comparatively slow. Some of the most ambitious plans for extensive automation of air traffic control functions initially envisage no facilities for intervention by either pilots or controllers in functions which hitherto they have been familiar with, and may even fail to inform either of them that the functions are taking place. These plans tend to be scaled down in the face of practical realities. Among the most pressing of these realities are the following: escalating costs, technical feasibility, underestimates of the complexity of the required software, difficult legal issues of responsibility, problems of human-machine matching, serious slippage in timescales during system procurement, training problems, the practical impossibility of foreseeing and allowing for every possible contingency, and lack of proof that tangible benefits of safety, capacity, or efficiency will accrue from the proposed automation. If it is concluded that some need will remain for the human to act as a back-up in non-standard situations, it can be very difficult to provide effective means to fulfil such a role through the human-machine interface without any mismatch of human and machine roles and requirements.

Historically the actual pace at which automation is being introduced into air traffic control is therefore often more ponderous than one might expect. When automation is successful, it is taken for granted by its users, to the extent that this could be construed as one of the criteria of successful automation. Users treat it as correct and as a basis for action, and are not forever trying to check or verify it. Successful examples from the past are the provision of identity and height information automatically on the labels of aircraft on radar displays, and indeed the very provision of these labels. The widespread replacement of paper flight progress strips with electronic versions of them wherever there is heavy commercial traffic has been pending for many years. Nevertheless, the traffic at some of the busiest air traffic control facilities is still being controlled using paper flight progress strips and amending them by hand. Plans for making this change are often well advanced, but the process of evolving a new air traffic control system, or even of making a major change in an existing system, is protracted at best and often characterised by slippage.

One explanation is the paramount requirement of safety. A revised system which would increase capacity and improve efficiency would still not be acceptable if it was demonstrably less safe. Yet air traffic control must evolve, so that each controller can handle more traffic, with the aid of automation. This combination of requirements leads to the progressive but cautious introduction of automation, with thorough pre-testing and the abandonment of some initially attractive alternatives which either raise unresolved issues of safety or do not actually deliver the significant increases in capacity that they promise.

The controllers who have to use the aids that are provided must find them acceptable and must be able to learn and understand how they function so that the usage of the aid is in accordance with the designer's intentions. This refers not only to aids which fulfil their original objectives successfully but also to aids which are applied by controllers to assist functions for which they were not intended and for which their use has not been legally sanctioned. This is partly a training issue, and it is essential to confirm that the correct usage of each aid is teachable. Controllers can be surprisingly inventive in finding uses for an aid which its designers never envisaged. It is necessary for controllers to know the full legitimate usage of each aid on completion of their training with it, but also necessary to curb their tendency to extend the planned applications of aids which they like.

A neglected human factors research issue has been the study of the formation of attitudes towards aids for the controller. Some are welcomed with enthusiasm and others condemned as unhelpful. For the latter it would be of real practical use to know before major resources are committed to developing them if adverse attitudes towards them are likely to be formed and whether more favourable attitudes could be engendered by alternative introductory training approaches. It would also be useful to know what the consequences are for performance if controllers have to use equipment which initially they do not like or do not trust. Successful aids are welcomed and accepted by their users, whose wishes have influenced their design and whose correct usage of them has been fostered by appropriate training. This thinking has become more orthodox and influential in recent years. It should help to ensure that the continued introduction of automation into air traffic control will achieve its expected benefits in the future by being used optimally by controllers who recognise that their skills continue to be valued and who appreciate that their jobs as air traffic controllers remain a source of pride and job satisfaction.

References

Boff, K. R. and Lincoln, J. E. (eds.) (1988), *Engineering Data Compendium: Human Perception and Performance,* Wright Patterson Air Force Base, Ohio: Armstrong Aerospace Medical Research Laboratory.

Cardosi, K. M. and Murphy, E. D. (eds.) (1995), *Human Factors in the Design and Evaluation of Air Traffic Control Systems,* Washington, D. C.: Federal Aviation Administration, Office of Aviation Research, Report DOT/FAA/RD-95/3.

Eurocontrol (1996), *Proceedings of the First Eurocontrol Human Factors Workshop: Cognitive Aspects in ATC, European Air Traffic Control Harmonisation and Integration Programme,* Report HUM.ET1.ST13.000-REP-01.

Federal Aviation Administration (1991), *The National Plan for Aviation Human Factors,* Washington, D. C.: Department of Transportation/ Federal Aviation Administration.

Hopkin, V. D. (1970), *Human Factors in the Ground Control of Aircraft,* Paris: N.A.T.O. Agardograph Number 142.

Hopkin, V. D. (1988), 'Air traffic control', in Wiener, E. L. and Nagel, D. C. (eds.), *Human Factors in Aviation,* pp.639-663.

Hopkin, V. D. (1994a), 'Colour on air traffic control displays', *Information Display, 10(1),* pp. 14-18.

Hopkin, V. D. (1994b), 'Optimizing human factors contributions', in Wise, J. A., Hopkin, V. D. and Garland, D. J. (eds.) *Human Factors Certification of Advanced Aviation Technologies,* Daytona Beach, FL.: Embry-Riddle Aeronautical University Press, pp. 3-8.

Hopkin, V. D. (1995), *Human Factors in Air Traffic Control,* London: Taylor & Francis.

Hopkin, V. D. (1997), 'Automation in air traffic control: recent advances and major issues', in Mouloua, M. and Koonce, J. M. (eds.) *Human-Automation Interaction: Research and Practice,* Mahwah, NJ: Lawrence Erlbaum, pp.250-257.

I.C.A.O. (1993), *Human Factors Digest No. 8: Human Factors in Air Traffic Control,* Montreal, Canada: International Civil Aviation Organization, Circular 241-AN/145.

MacLeod, I. S. and Taylor, R. M., Does human cognition allow human factors (HF) certification of advanced aircrew systems?', in Wise, J. A., Hopkin, V. D. and Garland, D. J. (eds.) *Human Factors Certification of Advanced Aviation Technologies,* Daytona Beach, Fl.: Embry-Riddle Aeronautical University Press, pp.163-186.

National Research Council, (1997), *Flight to the Future: Human Factors in Air Traffic Control,* Washington, D. C.: National Academy Press.

Reynolds, L.(1994) 'Colour for air traffic control displays', *Displays*, 15(4), pp. 215-225.

Salvendy, G. (ed.) (1987), *Handbook of Human Factors*, New York: Wiley.

Sanders, M. S. and McCormick, E. J. (eds.) (1993), *Human Factors in Engineering and Design*, New York: McGraw-Hill.

Stager, P. (1991), 'Error models for operating irregularities: implications for automation', in Wise, J. A., Hopkin, V. D. and Smith, M. L. (eds.) *Automation and Systems Issues in Air Traffic Control*, Berlin: Springer-Verlag, N.A.T.O. ASI Series F, Volume 73, pp.321-338.

Westrum, R. (1994) 'Is there a role for a "test controller" in the development of new ATC equipment?', in Wise, J. A., Hopkin, V. D. and Garland, D. J. (eds.) *Human Factors Certification of Advanced Aviation Technologies*, Daytona Beach, FL.: Embry-Riddle Aeronautical University Press, pp.221-228.

Wise, J. A. and Hopkin, V. D. (1997), 'Integrating human factors into the certification of systems', in Mouloua, M. and Koonce, J. M. (eds.) *Human-Automation Interaction: Research and Practice*, Mahwah, NJ.: Lawrence Erlbaum, pp.181-185.

Wise, J. A., Hopkin, V. D. and Garland, D. J. (eds.) (1994), *Human Factors Certification of Advanced Aviation Technologies*, Daytona Beach, FL.: Embry-Riddle Aeronautical University Press.

Wise, J. A., Hopkin, V. D. and Smith, M. J. (eds.) (1991), *Automation and Systems Issues in Air Traffic Control*, Berlin: Springer-Verlag, N.A.T.O. ASI Series F, Volume 73.

Wise, J.A., Hopkin, V.D. and Stager, P. (eds.) (1993), *Verification and Validation of Complex Systems: Human Factors Issues*, Berlin: Springer-Verlag, N.A.T.O. ASI Series F, Volume 110.

The author

Mr. V. David Hopkin, M.A., is a chartered psychologist and a Fellow of the Royal Institute of Navigation He is an independent Human Factors consultant. He formerly acted in this capacity full time to the United Kingdom Civil Aviation Authority and was a Senior Principal Psychologist at the Royal Air Force Institute of Aviation Medicine, Farnborough (now part of the DRA Farnborough), U.K.

He has carried out consultancy work for ICAO, Eurocontrol, FAA, CAA together with many international and national organizations and firms. He has lectured very extensively and now teaches regularly at Embry-Riddle Aeronautical University in Florida, U.S.A. Having published over 300 refereed publications, as well as being an author or joint editor of nine textbooks, his recent publication, in 1995, was *Human Factors in Air Traffic Control*, London: Taylor & Francis.

9 Training issues in Air Traffic Flow Management

John A. Wise, V. David Hopkin and Daniel J. Garland

Introduction

As the amount of air traffic increases worldwide and the fiscal constraints on the growth in air traffic systems become a limiting factor, the need for improved traffic flow management systems has become widely recognized throughout the world. As a result, practically every industrialized country is trying hard to improve the performance of its existing traffic flow management (TFM) system. The international community is actively promoting advanced computer tools that claim to enhance TFM system safety and performance. Most aviation agencies are actively developing and/or installing automated aids to be used in both operations and maintenance of TFM systems.

In the near future, these computer-based aids and other new technologies will revolutionize the international traffic flow management system and thus the traffic manager's job. Traffic flow management will evolve from the current 'hands-on' tactical control to strategic control and planning because of the capabilities of these automated systems. Consequently, there is a real concern in the TFM user community, that radical and progressive system changes may impose requirements on the traffic manager that are incompatible with the way the traffic managers have traditionally been trained.

The implementation of the new automated aids, while beneficial, will not resolve all the operational problems of TFM. In fact, new problems are inevitable. For example, while automated aids (e.g., expert systems, memory aids, decision aids) are intended to enhance a traffic manager's information processing capabilities, these aids may be in conflict with the traffic managers current cognitive models of TFM procedures and thus actually decrease the probability of effective decision making in the air traffic environment.

The cognitive requirements of any TFM system (automated or manual) necessarily involves the processing of dynamic and varying information.

Training issues in Air Traffic Flow Management

Cognitive processing of real-time and forecast flight, airspace, and weather data are crucial to virtually every aspect of a traffic manager's performance. It is essential for the traffic manager be able to use such information resources in such a way that effective problem solving and decision making can be performed when needed. The effectiveness with which the necessary information (e.g., flight data, airspace data, weather data) is processed, utilized, and remembered by the traffic flow manager depends in no small part on how, when, and where it is displayed, the transparency of the system interface, and the knowledge the manager has received though training and experience.

TFM system success also depends on the relationship between training, design, and selection. One can not discuss training without making assumptions about the other two. One should think of the three variables as the corners of a triangle as shown in fig. 9.1 If all three attributes were well understood and properly implemented, then a strong, reliable, and effective system will emerge. If any one of them is weak the entire system becomes unstable, unreliable, and thus ineffective. For example, it is impossible to implement an effective TFM training program without adequate knowledge of the personnel to be trained and the design of the TFM system.

Figure 9.1. TFM system success

While the idealized TFM system from a human factors engineering point-of-view is one that is so intuitive and easy to use, that anyone can operate it perfectly the first time without training, such an design for anything but a trivial application is not realistic. In any case, it is not sufficient to devise, design, and build traffic flow management workspaces within which all the required tasks can be performed efficiently, safely and in a timely manner. It is also essential that all the required human functions and roles are teachable, so that most, and preferably all, traffic managers are able to learn them easily, in order to minimize training costs. The trainability of TFM functions, procedures, and instructions therefore must be demonstrable. Preferably all TFM positions should be designed to support the process of learning them, to reinforce human memory and understanding, and to incorporate self-teaching aids (e.g., stand alone tutorials) wherever possible.

The nature of Traffic Flow Management

Air traffic control (ATC) and traffic flow management (TFM) together form a comprehensive system to carry out the responsibility for safe and efficient operation of national and international airspace. The TFM function is performed by traffic managers located at ATC facilities and TFM command centers. Traffic managers focus on a larger traffic picture organizing complex traffic flows through busy areas in the airspace and managing the volume of traffic into and out of congested airport areas. The primary responsibility of the traffic manager is to manage the flow of air traffic through the airspace while maximizing airspace efficiency and minimizing the impact of system constraints on air traffic facilities and airspace users.

To effectively accomplish TFM objectives there are several primary strategies. For example, (a) Metering – controlling the rate of traffic flow (miles/minutes in trail), (b) Rerouting – altering the distribution of traffic flows over a set of alternative routes or restructuring routes, and (c) Ground holding/stop – intentionally delaying/stopping an aircraft's take-off for a specified period of time. When strategies are executed, their effect on traffic is constantly monitored, evaluated and adjusted as necessary.

The picture of the airspace and traffic situation should support the traffic manager's activities to manage future traffic demand and capacity, principally scheduled air carrier operations, and airport acceptance rates. The traffic manager must be aware of traffic variables based on airline schedules, and other factors, such as weather or runway closures that may cause flights to divert to alternate airfields. Capacity, (i.e., runway acceptance rate) is also influenced by weather and runway closures. Choke points can also occur enroute as well as at departure and destination airports. Major changes in either volume or capacity that overload the system cause flight delays which equate to additional expense and passenger inconvenience (e.g., late arrivals, missed connections, etc.).

The traffic manager must consider the impact of flow restrictions in both a local and global sense. Restrictions or diversions may impact the traffic volume downstream at other locations. For example, a restriction that causes holding aircraft on the ground may also cause ground congestion that impedes acceptance of arriving aircraft.

Caveat

TFM procedures and policies tend to be very country dependent. The source of the traffic managers, the types of technology employed, the operational philosophy, the environmental constraints, the types and volume of traffic all impact the task performed by the traffic flow manager. As a result, it is impossible to give anything other than general guidelines for TFM training systems. This paper is primarily based on the work related to the U.S. traffic flow management system, with which the authors are most familiar, but care has been taken to try to address this topic in a way way which is fairly generic.

Basic cognitive issues

The cognitive requirements necessary for the proper development, maintenance, and enhancement of the requisite TFM skills basically involve the ability to process dynamically changing and interacting information within the operational and political constraints of the aviation system of which it is a component. These cognitive processes are fundamental to effective traffic management, and thus TFM system performance. These processes involve extracting, integrating, assessing, and acting upon task-relevant and usually time-sensitive information. Cognitive processing and integration of flight, airspace, and weather information is significant to every aspect of a traffic manager's performance. It is essential for the traffic manager to acquire the requisite information resources in such a way that they are available when needed.

Traffic flow management operations are exemplified by varied demands, information loads, as well as political and financial pressures. Because TFM operations are extremely time-sensitive the use of highly practiced automatic response patterns and the employment of the full range of cognitive capabilities are required to utilize the relevant data, information, and knowledge to effect a timely and correct response.

Superior TFM is a product of superior cognitive capabilities. The strength of an experienced traffic manager's capabilities is founded in domain-specific cognitive skills, e.g., knowing how task operations work and when to use them. The acquisition of this knowledge is an important step in understanding how TFM skills evolve and are maintained. Thus without effective training, there will be large differences in individual and thus system performance. Without effective training different traffic managers, even given identical information, can have dramatically different interpretations and conclusions. Thus, the decision of one manager might not be based on everything that another manager would assume it contained.

Because TFM performance depends on learning, performance should to some extent be explicable and predictable in terms of learning concepts and theories (assuming they are valid). The following assertions about traffic management skills and performance are made:

- Traffic flow management skills can be built, extended, developed, entrenched, and reinforced.

- Traffic flow management performance will be improved through the acquisition of appropriate skills, by the development of expertise, by a more extensive knowledge base, and by improved accessibility of that knowledge base.

- Traffic flow management skills can be fallible, incomplete, distorted, subject to error, or forgotten.

- The quality of traffic management skills can atrophy with disuse, so practice will have a crucial role in maintaining that capability. For example highly automated systems have a requirement to keep cognitive skills current, otherwise the managers will lose them.

- Traffic managers will be subject to the formation of habits and thus may be 1) resistant to new evidence that appears to contradicts previous beliefs, 2) biased in the choice of evidence that is relevant to it, and 3) over-influenced by particular memories which may be treated as more relevant than they are.

- Learned meanings, which form an intrinsic part of perceptual processes, will be resistant to the recognition and correction of any errors that they embody (e.g., when a three-digit number has been recognized as a flight level, it may be difficult to acknowledge that it is actually a speed or a heading).

- Traffic flow management skills will be influenced by training, by what is taught, by how it is taught, and by the relevance of what is taught to what is needed.

- Traffic flow management skills will be influenced by motivation, interests, job satisfaction, self-esteem, and the esteem of others.

- Traffic flow management skills can be influenced by attentional factors.

- Traffic flow management skills will not necessarily incorporate all stimuli, but only those that are perceived as meaningful, and items that seem meaningless will be ignored (no matter how important they may actually be).

Given the above, traffic flow management training has to take account of the basic cognitive capabilities of people, how they think, how they decide, how they understand, and how they remember. Jobs and tasks must also be designed within these capabilities and training must be devised to maximize their effective use. Traffic managers need to be able to use their cognitive capabilities well and sensibly, but also in a way which they recognize as worthwhile and satisfying.

Mental models

Each traffic manager has an internal mental representation (or model) of the air traffic situation that determines the response to information displays that are intended to provide a representation of the real world. The way that an individual traffic manager perceives information and the resulting decisions, and actions, is determined by the training process, which in turn facilitates development of individual mental models. The display interface needs to be designed to be compatible with the traffic managers mental model, so that symbols, alerts and messages are clear, unambiguous and support an accurate interpretation of the situation. It is very important that the display designer understand the traffic manager's mental models when selecting display methods.

The importance of mental models to human-computer interface design, especially as it relates to process control may be summarized as follows. A display, process, or system must be compatible with the operator's internal representation, or mental model, and should enable the construction and refinement of mental models to enhance performance. This, of course, implies that interface design can influence the formation of the user's mental mode and, in fact, is especially true in supervisory control and automated systems where the operator is in indirect rather than direct control of the process or system.

Experienced air traffic managers' knowledge is hierarchically organized and consequently, displays and training programs should be organized likewise; emphasizing system functions at high levels and system structure at low levels (Murphy and Mitchell, 1984). This is relevant to the design of primary situation displays that allow air traffic managers to maintain the "big picture" and the ancillary displays that support decision making and resolution analysis. The traffic manager should have the ability to navigate efficiently through displays presenting different levels of information. Displays should also support the ability of the traffic manager to maintain situation awareness as well as his or her position within the hierarchy and not lose sight of the primary task of monitoring the dynamically evolving ATC traffic picture within his or her sector.

There are many possibilities for future visual codings for flows of aircraft. For example, one possibility is to use dynamic density displays to convey the general state on the overall airspace system, with the ability to get detailed information on potential problem areas using more detailed displays such as, real-time display of spare air route capacity displays, or time-based/predictive displays. Whatever the display chosen, the coded information should be applicable directly to specific traffic manager tasks and should require neither recoding nor computations by the traffic manager before it can be used.

Suitable codings to represent flow patterns need to be devised and proved. The cognitive aspects of the displays should be investigated as well, for instance, determining which displays are easier to hold in short term memory. Information should never be displayed just because it is technologically feasible to do so. In addition, time and resources permitting, a systems design approach with empirical testing should be an integral part of the design of any new display. In designing new displays one must be aware of the potential for additional data reducing overall performance by introducing extraneous data which can mask the relevant information or lead to 'information overload'.

Basic research on visual codings (e.g., Bennett, Toms, & Woods, 1993; Casali & Gaylin, 1988; Sander & McCormick, 1993) and new display technologies (e.g., Woods, Wise, & Hanes, 1981, Hopkin, 1994) need to be reviewed and considered. Furthermore, coding techniques used in other fields such as cartography (e.g., McCleary, Jenks, & Ellis, 1993; MacEachren, 1995) and process plants (e.g., Umbers & Collier, 1990) have a strong potential for being useful sources for ideas and guidance.

In addition, with the advent of 'free-flight', it is important to remember that as tasks and functions change, the information needed to perform them normally changes too (Garland, Hopkin, & Muller, 1996). Air traffic problems can quickly grow from regional to national in scope, so it is of the utmost importance that useful information is presented to air traffic managers in a coherent fashion.

Information presentation issues

Several computer-based tools are used to support TFM tasks. In the United States one of the often used aids by traffic managers is the aircraft situation display (ASD) (see Figure 9.2). A variety of visual codings of traffic flow management information categories have evolved on the ASD with different users in different locations. The ASD graphically displays current aircraft positions on a national scale superimposed on maps of geographical boundaries and/or NAS facilities. It can display a multitude of information (e.g., data tags, weather data, low, high, and super high sectors, victor airways, jet routes, etc.). The ASD user can also select numerous methods of filtering to highlight groups of aircraft since the system can display all known flights simultaneously, or filter out all but a selected group of flights. As the primary tool used to plan and monitor TFM initiatives, the ASD offers a large number of options for the display of different pieces of data.

The experience of ATC radar controllers appears to have been a major factor in ASD design. For example, displays show location, heading and data block (optional) of each aircraft. A presentation like this may be acceptable to the traffic manager because it looks familiar. That familiarity even extends to the stated preference for a black background, the same as the radar screen. Colour aircraft icons are shown in the position relayed from radar. Filters can be selected to show only aircraft associated with any specific location. Colour is selectable to indicate several different conditions, e.g., different colours for arriving and departing aircraft, specific aircraft types, etc.

Figure 9.2 Aircraft situation display (ASD)

The method of presentation may not be the most effective because the role of the air traffic manager differs, significantly, from the role of the radar controller. Consequently, information presentation in a format more suited to situation analysis and development of alternatives may be more effective. Communications and coordination with the various airports and enroute sectors involved in a traffic situation are essential activities of the traffic manager. Information presentation should be tailored to support the specific traffic management decisions and actions in a timely manner to accomplish the desired objective.

The fact that the traffic manager's job is very different from the controllers' job has not been reflected in an interface that is tailored to support the traffic manager's activities and decision. A significant question is whether the traffic manager at the national level needs to be able to see individual aircraft, or whether some other presentation would be more effective. Alternative display concepts should be developed and evaluated with input from experienced traffic managers.

Information presentation and training requirements differ at the different operating levels; e.g., national, regional, and local. Each level has different goals that influence the areas of interest and the level of detail that is useful. At the regional level, the goal may be to manage the traffic flow to fully use the available capacity at the current acceptance rate, and balance workload between controllers. At the national level, the goals may be more strategic, that is to minimize the impact of delays on air traffic, avoid holdings, and balance the traffic load. Locations of individual aircraft and high traffic volume areas is useful information for the traffic managers at regional facilities.

Information displays and training support are needed to provide relevant traffic management analysis and decision making support. Presentation of individual aircraft symbols appears to have little value compared to the extensive computation power available. Notification that there is a situation that requires a change of the traffic management plan usually comes from local controllers who determine the need for action based on an event, or from weather observations. Ideally, TFM actions should be proactive and preventative in nature, not reactionary. In the United States, the ASD provides information that is useful for understanding the scope of the traffic situation and for coordinating proposed actions between remote facilities.

Situation analysis is an essential element of traffic management and requires the skill to quickly identify the traffic management strategy that resolves the current restriction with the least impact on other traffic. Computer-based tools need to provide information to facilitate situation analysis and decision making. These should be configured to support analysis and decision making by providing acceptable alternatives.

Although TFM tasks are generally strategic and deal with groups or flows of aircraft, the information provided on visual displays for these strategic tasks is mostly tactical, and takes the form of data about single aircraft rather than about groups or aircraft flows (Garland, Hopkin, & Muller, 1996). Frequently the most used tools enacted by traffic managers are those normally associated with controlling 'live' traffic, such as data tags, when in fact the traffic manager has no operational control over (or link to) individual aircraft. There appear to be no coding of data in common use that represent groups of aircraft or flows of aircraft traffic directly, with the result that data about groups and flows have to be gleaned from impressions of traffic patterns derived from data on single aircraft and from the "current situation" information, requiring the traffic management specialist to extrapolate from the present to the future. This does not facilitate or encourage strategic thinking and planning though most of the tasks of the traffic management specialist are of a predictive nature.

Training issues in Air Traffic Flow Management

The codings, and particularly the flexibility of the codings on TFM displays, needs further development. Though the stated aims of TFM are largely strategic, the coding remains almost exclusively tactical. More flexible codings of traffic flows need to be developed. Codings are needed that discriminate visually between all the main parameters by which traffic flows are defined for traffic flow management purposes.

Coding flows

Future TFM displays will likely need to contain strategic information, by presenting groups or flows of aircraft. There are numerous possible visual codings for displaying traffic density and flow. One possible way of doing this would be to, through the use of mathematical algorithms, create density plots (as an example see fig. 9.3). Dynamic density is commonly thought of as consisting of traffic density, complexity of traffic flows, and separation standards, this involves not only the number of aircraft, but their pattern of movement in relation to each other. Research is needed to investigate what the various levels of density should represent, and how many levels should be used, likely this will depend on several factors (e.g., area of the country, time of day, altitude range being considered).

Figure 9.3. Dynamic density display

In addition to the more obvious codable characteristics of flows (location, direction of flows, distance/time, airline, flight level), there are numerous characteristics which will likely be much more difficult to code in intuitive ways. For example, a single flow may split into more than one flow and more than one flow may amalgamate into a single flow. Furthermore, both splitting and amalgamation may occur at the same location and time, or flows may cross without splitting or amalgamating (Hopkin, 1995). The best way to code this is far from clear. It may also be important to traffic managers tasks to represent traffic flows in various ways, such as maximum traffic flow, peak traffic flow (including the duration of this), average traffic flow, actual traffic flow, and predicted traffic flow, while making sure that these codings are not easily confused (Hopkin, 1995).

The essential information for traffic flow management concerns where there is or will be a shortfall in capacity and where there is spare capacity (Garland, Hopkin, & Muller, 1996). Spare capacity should, if possible, be presented directly rather than requiring computation by subtracting current capacity from maximum capacity.

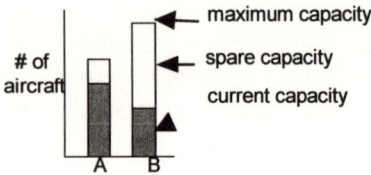

Figure 9.4 Capacity\ratio graph

One way to do this is with stacked bar graphs. Stacked bar graphs (see figure 3), whether used to show a snap-shot in time to compare resources (A & B representing separate jet routes, fixes, airports, etc.) or to show the trend of a single resource across time (A & B representing separate points in time), have the advantage of presenting a lot of information succinctly and clearly. Not only are the ratios of current/maximum capacity, and spare/maximum capacity given, the absolute number of aircraft involved can be compared either across time or across resources, depending on how it is used. Additionally, if the x-axis is time, the trends of current, spare, and maximum capacity can all be easily derived.

Codings must make it clear what any spare capacity refers to. It may be necessary to do this in two distinct ways. One would refer to time interval so that, for example, a given number of aircraft could be added to a route in a fifteen minute period. The other refers to geographical distance and it may be necessary to use some form of tags to mark the geographical segments to which the spare capacity figures refer, or to brighten or otherwise highlight the symbols or labels of the positions on the route between which the spare flow capacity figures apply (Hopkin, 1995).

For some tasks, it is necessary to know the flow rate in terms of the number of aircraft passing through a fix or the number of aircraft flying a route segment within a given period of time. In order to accomplish certain tasks, it may also be necessary to know how this rate compares with other previous or predicted rates. This involves the capability to make comparisons on the displays to show changes in flow rates, the magnitude of changes in flow rates, and differences between changes in flow rates. In many cases it might suffice to provide information that shows trends rather than very precise quantities (Hopkin, 1996).

Figure 4 shows a route segment with aircraft flying in both directions, with the width of the segment depicting maximum capacity, thereby allowing the depiction of spare capacity and surpluses of capacity. Surpluses are shown by the darker grey areas overrunning the rectangle which represents the air route.

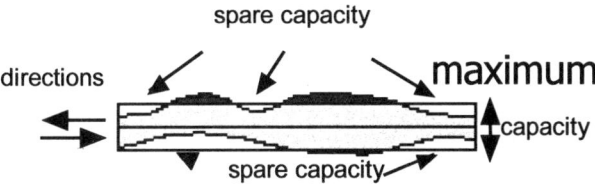

Figure 9.5. Route density display

Selection or background issues

The cognitive skills that appear to be most relevant for traffic flow management tasks may not be among those for which they were used to originally select operational air traffic controllers or which they have learned as controllers to hold as valuable. Several issues related to selection and training arise from this. One concern is the relevance of current air traffic control jobs to TFM tasks, particularly in terms of their relative cognitive requirements. Some existing controller jobs, such as that of the oceanic planner, might require attributes similar to those that seem relevant to TFM tasks, but more detailed analysis of the cognitive aspects of such jobs would be needed to establish how far their desirable attributes overlap. An associated issue is whether the possession of attributes favoring strategic cognitive thinking is relatively innate and should therefore be measured as part of the procedure for selecting traffic managers and perhaps controllers, or is it mainly learned and therefore primarily a product of suitable training.

While the ability to think strategically, rather than tactically, in traffic flow management positions may be extremely important, the opposite problem must not be ignored. In those systems where personnel perform TFM roles for limited periods and then ultimately resume air traffic control positions (a task that requires tactical thinking) there will also be need for persons who can unlearn TFM strategic thinking and relearn ATC tactical thinking. Thus, training will be needed to help the manager think strategically in the TFM role, and then comparable retraining may be needed to allow a manager to think tactically again on returning to an air traffic control position.

It is possible that current selection procedures for controllers reject many candidates with attributes which fit them well for traffic flow management roles. Related issues are whether ATC selection procedures should be extended to take account of this, and whether the potential ability to fulfill TFM roles well could be identified from any measurements taken as part of the present selection procedure for controllers. A further possibility is that selection for TFM positions should employ a specially developed, standardized and formalized selection procedure to choose the most suitable candidates from the existing controller population. Whatever procedures are adopted to allocate controllers to TFM positions, some allowance must be made to refine and improve those procedures and to measure and increase their validity as more evidence about them accumulates.

As long as there is a sufficient supply of candidates for traffic flow management, it may be beneficial to develop a selection procedure to identify those with the most potential ability for TFM tasks. It is prudent to do this in any event, to be able to allocate to TFM those who could do the job best because the requisite abilities have been identified and quantitative tests of those abilities have been devised and validated. This requires human factors work to build and prove an appropriate selection procedure and to ensure that it is independent of training.

During the course of selecting air traffic controllers, a number of tests could be administered with the intention of placing the individual in the proper position. These tests should be designed to indicate the potential for an individual to cope with strategic tasks required of the traffic manager. Traffic managers, both experienced and in-training, often operate in a manner consistent with air traffic control. There is a need for critical examination of the nature of the traffic manager position in the current and future systems, not only for reasons of role clarification and performance measurement, but perhaps more importantly for supporting the proper training when transitioning from the tactical to the strategic environment.

The opinion is widely held that traffic management performance improves with experience. Although eventually you reach a threshold, after which any additional experience does not necessarily mean an increase in performance. If this is true, it is desirable to discover why this occurs and in what specific ways, with a view to incorporating into the formal traffic management training the knowledge that is currently gained from experience. Experience teaches what is practical, what is successful, what works out according to plan and what does not, what the practical options are that are worth considering, and how flow problems can be categorized and classified as a basis for judging which previous solutions can be adopted again and which cannot because certain crucial factors have changed. It seems worth expending effort to ascertain what could profitably be taught formally that is not taught now, what else traffic managers actually learn from experience, and whether this further learning is teachable.

Basic training issues

Training is a matter of learning, understanding, and remembering. It relates what the traffic manager already knows to the information that the system provides about current and pending traffic. It relates information which the system presents to the traffic manager automatically, to information which the traffic manager must remember unaided and provides guidance on how human memory can be strengthened and become more reliable. Training also relates the principles for learning and displaying TFM information, to the capabilities and limitations of human information processing and understanding. The aim is to make the best use of human strengths and capabilities and to overcome or circumvent human inadequacies or limitations, particularly in relation to knowledge, skill, information processing, understanding, memory, and workload.

The objective of TFM training must be to ensure that traffic flow managers possess the required skills, knowledge, and experience required to perform their duties safely and efficiently, and to meet national and international standards. Therefore, at a general level TFM training must provide the traffic manager with the ability to:

- Operate at the appropriate functional level

- Work effectively with a team with disparate goals

- Plan ahead

- Understand operational situations

- Acquire, sort, and utilize relevant information

- Make timely and appropriate decisions

- Implement decisions system wide

- Ensure compliance with decisions

In introducing systems changes, for whatever reason, it is vital to determine what new knowledge (if any) the traffic manager must acquire and to show that it can be taught and learned. One can deduce from envisaged tasks what the traffic manager must know. This can be compared to current knowledge requirements to determine what the content of the new training must include. One then needs to determine if it can be taught or if traffic managers can learn it. New forms of automated assistance must be teachable. If they are not, their expected benefits will not materialize and new forms of human error will probably arise.

The content of what is taught, the sequence of which items are taught, the pace of teaching and the amount of reinforcement and rehearsal of taught TFM information should all be established according to known learning principles. Knowledge of results and of progress is essential for successful learning. The efficiency of learning depends on teaching methods, content and presentation of material, attributes and motivation of the student and on whether the instruction is provided by a human, computer, or both. It also depends on whether the instruction is theoretical or practical, general or specific.

The traffic manager needs to know and understand at a minimum:

- How traffic flow management is conducted.

- The applicable regulations, procedures, and instructions.

- The tasks to be accomplished.

- The way in which work of various traffic management personnel fits together cohesively so as to provide support and not to impede each other.

- Aircraft performance characteristics and preferred manœuvring.

- Influences on flights and routes, such as weather, restricted airspace, noise abatement, etc.

- Every tool provided within the workspace, and how, when, and why to use each.

- The forms and methods of communications in the system.

- The meaning of all presented information.

- What changes or signs could denote system degradations or failures.

- Human factors issues in TFM (e.g., communication and negotiations).

- Learning and understanding all the rules, regulations, procedures, instructions, scheduling, planning, and practices relevant to the efficient conduct of TFM.

- Procedures for liaison and coordination with colleagues, ATC, and system users.

- Recognition, prevention, and mitigation of operator-induced, design-induced, and system-induced errors.

- Matching the computer assistance to the traffic manager so that any errors are detectable, prevented, and corrected.

- Verification of the training progress of each traffic flow management student by impartial assessments that are accepted as fair by all.

- Acquisition of knowledge about professional attitudes and practices with TFM.

- Acceptance of the professional standards that prevail and the personal motivation always to attain and exceed those standards.

Training should follow not only established pedagogical theories, but also human factors procedures and practices. It should be flexible enough to be adaptable to the needs of individual traffic managers. It should incorporate a basic understanding of human factors so the traffic managers have some insight into their own capabilities and limitations, particularly with regard to possible human errors and mistakes. Traffic managers should know enough to be able to select the most appropriate automated aids to improve their task performance and efficiency, especially in display options.

There are some signs that what is taught about traffic flow management may sometimes have a tenuous connection with the practice of TFM. This may be a consequence of the facilities available for teaching, particularly if the provision of training facilities lags behind the provision of the corresponding operational equipment, as it often does. Sometimes the training becomes less useful if changes in the content, methods, and flexibility of training are out of step with system changes. Although it is premature to make specific generalizations about the outcome of training, it seems intrinsically unlikely that everyone who undergoes training in TFM possesses all the talents needed for it. It can waste training resources to continue training everyone regardless of their progress, but there may be no practical alternative in the absence of any independent, validated and objective criteria that can be applied early during training to identify those least capable of becoming proficient in traffic flow management in order to discontinue their training.

If it becomes desirable or necessary to shorten training, the following techniques might be considered to achieve this end:

• More computer assistance for routine functions

• More direct machine guidance on options, present state, and what has already been achieved

• More information about the consequences of the choices offered

• Clearer criteria on the choice of options

• More self-evident labeling of options

• Better evidence on the criteria for revising the presented information and on the frequency with which it is updated

• Better evidence on the integrity of data, particularly if any data can be misleading because of a lag in up-dating them

Automation impacts

A human factors problem that has arisen in air traffic control contexts will recur in traffic flow management. The computer assistance available seem more suited to assist individual traffic managers than teams of traffic managers. This computer assistance will tend to steer key TFM activities towards human-machine dialogues conducted by a single person through a human-machine interface designed to be used by only one person. Problems therefore may arise in informing other traffic managers about the activities and progress of their colleagues, in order to coordinate work. How much does each traffic manager need to know about their colleagues concurrent activities while doing his or her own tasks? How can training be utilized to counter this potential problem?

The rapid advancement in technology and its impact on the future TFM system design will need continual examination. Human factors specialists are just beginning to understand the problems associated with advanced levels of automation. Industries (e.g., nuclear power, aviation) have shown that highly automated systems require new training paradigms; thus the future TFM training program will need to be developed in light of recent empirical findings. For example, previous research in the air carrier environment has found that training practices have not kept up with advancing technologies, particularly in automated cockpits. As a result, the major airlines experienced tremendous turnover rates with pilots going through initial 'glass cockpit' certification, constituting unforeseen training costs.

To alleviate some of the initial training problems, different methodologies will be required to train traffic managers. Perhaps the most critical approach would be 'error training'. Such training would help the traffic managers develop a deeper understanding of the automation logic and form an accurate mental model of the representative system. This type of training would encourage an understanding of 'why' and 'how' the automation works. In addition, training on how to conduct monitoring without slipping into boredom or complacency must also be given high priority.

Research has also shown that automated aids in the glass cockpit environment (e.g., flight management system) require an inordinate amount of training to learn and operate at peak effectiveness. The same could be expected/should be anticipated for automated aids in the future TFM system. Regulations, procedures, and proficiency standards will need to be established and developed for a new training system that would ensure at least a minimum of training in specific areas is addressed to allow for adequate knowledge and safety.

The issue as to if or when traffic managers transfer back to previous duties as an air traffic controller also raises an interesting question. Will they be able to perform the duties at the required proficiency level(s)? Will they experience some skill deterioration because of the reliance upon highly automated assistance? If there is a skill loss, it is likely to be cognitive because of the nature of the job. This dilemma appears to be innate with automated systems and has been expressed as a major concern when pilots revert back to non-automated aircraft after flying the glass cockpit. Although there are no empirical data to support such allegations, several observations and numerous reports by pilots emphasize the problem.

Training and system changes

Change in any aspect of the TFM system that affects what the individual traffic manager should do or needs to know, such as a new form of automated assistance, should normally be associated with a careful redefinition of all the consequent changes in the traffic manager's knowledge, skills and procedures that will be needed. If the system has been changed, what the traffic manager needs to know about it will also have changed. Appropriate retraining should then be provided before the traffic manager encounters these changes while managing real traffic. The benefits of any changes to the TFM system that affect the traffic manager will be gained fully only if the corresponding changes in the traffic manager's knowledge and skills to match the system have been instilled through appropriate retraining beforehand. It should be normal for traffic managers to have refresher training regularly, during which knowledge and skills are practiced and verified and changes are introduced if they are needed. Wherever possible, the changes made in TFM systems should allow the existing knowledge and skills of traffic managers to remain applicable in the new system.

The introduction of new TFM tools to the field needs to be accompanied by clear and distinct training on their purpose, potential benefits, capabilities, limitations, and usage. Introducing new tools to the field without the necessary orientation and training can be problematic. If, traffic flow management tools are fielded without the opportunity to gain any significant experience with them, and the intended operational functions are not clearly defined, then it will be left to the traffic managers in the field to explore, or not to explore, its capabilities and to determine how it may be of use to them. Too often new tools are implemented and not accompanied by proper documentation and training as to its objectives or purposes, by instructions on its usage or by information on its applications.

This method of introducing equipment has some significant implications. It can become quite time-consuming to discover the equipment capabilities. It is highly probable that some of its possible applications will never be discovered. The equipment may be employed for purposes for which it was never intended. Very different applications of the tools may evolve at different locations, and there may be little standardization or commonalty in their use. A further implication is that people at other facilities may presume wrongly that others use the tools as they do, and draw unwarranted conclusions about what others must know. If the tools are being applied to different tasks in different places and the tasks are selected on a rather arbitrary basis, it becomes impossible to optimize the tools for all their functions since the latter are so diverse. The original intentions in providing the TFM tools have been formulated, and these need to be conveyed to achieve more uniformity and consistency in their applications

Continual change

As experience in using the traffic flow management system is gained, improvements in what is achievable should gradually accrue. These improvements will affect many aspects of the TFM system. Experience should provide better evidence on the precise circumstances when the TFM system is most essential and most beneficial, on the kinds of planning and intervention that are most effective, on the range of options that are worth considering, on the particular conditions when previously satisfactory solutions can be adopted again or when new solutions must be tried, on the optimum timing for any TFM system revisions, and on numerous further TFM system operations. The extent to which improvements from experience can occur depends on the forms of feedback available about the success of previous TFM system interventions and programs. Learning from experience cannot take place without feedback that is appropriate in its form, content, timing, and level of detail.

Positive forms of feedback, which can demonstrate to the traffic manager better solutions than those actually adopted or can enable the traffic manager to deduce better solutions, can be very helpful in enabling the human to learn from experience. Equally important, but comparatively neglected, are more negative forms of feedback, it can demonstrate that the solutions adopted, although they might have seemed far from ideal at the time or in retrospect, were in fact the best available because no greater improvements could possibly have been achieved by any alternative solutions. A related issue is how to apply models of the system to derive such forms of feedback.

A vital aspect of training is knowing when to stop, i.e., knowing when no further improvements can be expected from further experience. Such judgements depend on the derivation of valid and independent external criteria to establish what is attainable. System models may be capable of providing suitable criteria for this purpose.

All systems with the characteristics of the TFM system have the potential to seem unnecessarily alarmist. They can convey the impression beforehand that a problem will be more severe than it actually proves to be, or that a problem requires traffic management intervention when it can, in fact, be resolved satisfactorily by controllers using more tactical short-term strategies. One criterion for assessing the traffic flow management system, and an essential source of baseline data to make comparisons and to quantify its achievements, depends on a facility to describe in the same terms both what would have occurred without any traffic management action and what actually did occur with traffic management action. Such comparisons can then be used to vindicate the traffic management actions, to quantify its achievements, and to illustrate the kinds of solutions that yield the largest or most beneficial improvements.

A further possible role of system models, which is valuable in its own right and as a support for human factors work, is their usage to predict the outcome if available options that were discarded had in fact been implemented. Would they have been better than the option(s) actually chosen, and what human factors problems would have arisen in implementing them?

The above topic exemplifies the wider issue of how best to integrate different techniques, particularly real-time and fast-time techniques. Ideally, fast-time modelling techniques can identify combinations of circumstances requiring supplementary real time human factors evidence in order to improve the content and validity of the model. Also, real-time human factors studies can reveal anomalies and combinations of circumstances that present difficulties which appear to require exploration and explanation through the kinds of systematic manipulation of controlled variables for which models are well suited. Using these techniques to complement each other would strengthen both of them and offer further advantages, but their effective integration has always been more difficult to accomplish in practice than in principle.

Conclusions

If TFM systems are ever to reach their goal of improving traffic flow and reducing unnecessary delays, training systems that allow traffic managers use of the decision support technology and other tools will be necessary. These training systems need to be well designed based on established learning theory and the real world needs of day-to-day TFM operations. The purpose of this paper has been to identify basic training challenges in current and 'next generation' traffic flow management systems. Because the practice of TFM varies so dramatically across countries, this paper has only tried to identify the general issues that are relevant to the design and evaluation of training systems for traffic flow management systems. Where possible, an attempt has been made to identify training issues relevant to some of the more common problems (e.g., leaving TFM to return to ATC).

There may be some places outside of TFM that could provide the training system designer with some insight into effective training approaches. For example, high level military training programs that attempt to provide flag level staff with the ability to think and act strategically even when they are under high levels of stress. 'War games' used by the military and business simulations used in business graduate schools could probably be modified and successfully applied to the TFM environment.

Effective training systems can have a substantial positive impact on the world's air traffic. It is important that everyone involved in TFM actively involve themselves in assuring that the systems utilized do in fact support TFM system goals.

References

Bennett, K. B., Toms, M. L., & Woods, D. D.. (1993). Emergent features and graphical elements: Designing more effective configural displays. *Human Factors, 35*(1), 71-97.

Casali, J. G., & Gaylin, K. B. (1988). Selected graph design variables in four interpretation tasks: a microcomputer-based pilot study. *Behaviour and Information Technology, 7*(1), 31-49.

Garland, D. J., Hopkin, V. D., & Muller, J. K. (1996). *The Air Traffic Control System Command Center's Operational Workplace: Volume II - Human Factors Recommendations* (Rep. No. CAAR-15418-96-101). Daytona Beach, FL: Embry-Riddle Aeronautical University.

Hopkin, V. D. (1996). *Human Factors Issues Related to Traffic Flow Management, Including Some Arising from Free Flight.* (Rep. No. CAAR-15418-96-102). Daytona Beach, FL: Embry-Riddle Aeronautical University.

Hopkin, V. D. (1995). *Traffic Flow Management, Laboratory Research Projects. Codings for Flows of Traffic.* (Rep. No. CAAR-15418-95-105). Daytona Beach, FL: Embry-Riddle Aeronautical University.

Hopkin, V. D. (1994). Color on air-traffic-control displays. *Information Display, 10*, 14-18.

MacEachren, A. M. (1995). *How Maps Work: Representation, Visualization & Design.* New York: Guilford Press.

McCleary, G. F., Jr., Jenks, G. F., & Ellis, S. R. (1993). Cartography and map displays. In S.R. Ellis & M.K. Kaiser (Eds.), *Pictorial communication in virtual and real environments* (2nd ed., pp. 76-96). Washington, DC: Taylor & Francis.

Murphy, E., & Mitchell, C. (1984). Cognitive attributes to guide display design in automated command and control systems. In *Proceedings of the Human Factors Society 28th Annual Meeting,* pp. 418-422.

Sanders, M. S., and McCormick, E. J. (1993). *Human Factors in Engineering and Design* (7th ed.). New York: McGraw-Hill.

Umbers, I. G., Collier, G. D. (1990). Coding techniques for process plant VDU formats. *Applied Ergonomics, 21*(3), 187-198.

Woods, D. D., Wise, J.A & Hanes, L.F (1981). An evaluation of nuclear power plant safety parameter display systems. *Proceedings of the 25th Annual Meeting of the Human Factors Society.* Rochester, NY,October 1981.

The authors

Professors John A. Wise, V. David Hopkin, and Daniel J. Garland are all active in teaching and studying Human Factors and Training with respect to aviation in the Department of Human Factors and Systems at Embry-Riddle Aeronautical University, Daytona Beach, Florida, USA. Besides many publications they have been responsible for organizing international workshops which have addressed various topical issues in the Human Factors field for aviation.

Acknowledgements

This paper is based on research sponsored by the U.S. Federal Aviation Administration under contract 94-G-001. The opinions expressed are solely those of the authors, and do not necessarily reflect those of the sponsor.

Part 3

MANAGING THE AVIATION SYSTEM

10 The Air Traffic Management System – present and future

Vince Galotti

This chapter describes the shortcomings of present air traffic control systems and the changes anticipated for the future to reduce or remove these constraints.

The description, as well as the goals and objectives of the future air traffic management (ATM) system emanate from a concept of the future aviation infrastructure, originally known as the Future Air Navigation Systems (FANS) Concept. The concept, now referred to as 'communications, navigation, and surveillance/air traffic management (CNS/ATM) systems', was formally endorsed by the world-wide civil aviation community at the Tenth Air Navigation Conference, held in Montreal in 1991 (ICAO, 1991 & 1993).

CNS/ATM is the blueprint for the future aviation system developed to accommodate the needs of the air transport industry. The ultimate goal is a seamless, global ATM system that will enable aircraft operators to meet their planned times of departure and arrival and adhere to their preferred flight profiles with minimum constraints and without compromising agreed to levels of safety.

Air traffic control today: an interdependent system

The infrastructure that guides, separates and coordinates aircraft movement is termed the air traffic control (ATC) system, which is a part of air traffic services (ATS). ATS are provided by ground facilities, usually operated by national authorities and, in some cases, by international air traffic organizations composed of members of several nations (e.g., the European Organization for the Safety of Air Navigation (EUROCONTROL)). However, increasingly, States are transferring the operational service to autonomous organizations.

Developing the Future Aviation System

ATS is a generic term meaning variously, flight information service, alerting service, air traffic advisory service and air traffic control (ATC) service which itself encompasses area control service for enroute traffic, approach control service for arriving and departing traffic, and aerodrome control service for traffic arriving, departing and manoeuvring at the airport. ATC service by far, accounts for the greatest percentage of ATS on a global basis, and is provided for the purpose of preventing collisions between aircraft and between aircraft and obstructions on the manoeuvring area, and of expediting and maintaining an orderly flow of air traffic.

As traffic volumes grow, the demands on the air traffic services (ATS) provider in an airspace increase. For given separation standards, the number of flights unable to follow optimum flight paths increases. This creates pressure to upgrade the level of ATS.

Shortcomings and limitations of the conventional ATC system

Airport Surface Movement Area

The ground control of aircraft is conducted through radar or visual means. Automation to support surface movement guidance and control systems (SMGCS) of aircraft and vehicles is lacking, and many major airports operate in near gridlock conditions during periods of peak demand. There is also a lack of adequate co-ordination between ATC and ramp and taxi areas, necessary for the standardization and harmonization of gate-to-gate operations envisaged in future ATM systems. In low-visibility conditions, movements are severely restricted and there is increased risk of runway incursion or violation of instrument landing system (ILS) critical or sensitive areas (IATA, 1995).

Terminal control operations during climb and descent

Although terminal airspace usually is provided with better surveillance and communications capabilities than enroute airspace, it differs from en-route airspace primarily because of the higher traffic densities and greater complexity of traffic flow. Both arriving and departing aircraft share the terminal airspace, and also aircraft having widely differing performance characteristics operate to the same or closely spaced runways. Current separation requirements often prevent full use of available capacity at the busiest airports. Automation to manage departures and arrivals efficiently is, in most cases, not available and on-board automation is therefore under-utilized. Published arrival and departure routes are inflexible and result in indirect routings. Strict noise abatement and environmental procedures impose further restrictions on terminal area operations.

Enroute operations

The existing ATS route structure often involves mileage penalties compared to the most economic routes which may be great circle routes, but which also takes into account wind, temperature and other factors such as weight of the aircraft, charges and safety. This often results in concentration of traffic flows at major intersections, which can lead to a reduction in the number of optimum flight levels being available.

A lack in uniformity of ATC procedures and separation minima around the world, due to differences and limitations in ATC capabilities, places additional constraints on the aircraft operators. Furthermore, aircraft are often unable to take advantage of advanced on-board capabilities because the ATC system is unable to support their use.

A lack of co-ordination among States in the development of ground ATC systems has resulted in additional problems. Examples include inconsistent separation standards in radar and non-radar airspace and operation at less than optimum flight levels in oceanic airspace due to communication deficiencies.

The future Air Traffic Management System

The four main elements of CNS/ATM systems are summarized below and are dealt with in detail in the ICAO Global Air Navigation Plan for CNS/ATM Systems (ICAO, 1998).

Communications

In CNS/ATM systems, the transmission of voice will, initially, continue to take place over existing VHF channels; however, these same VHF channels will increasingly be used to transmit digital data. Satellite data and voice communications, capable of global coverage, are also being introduced along with data transmission over HF channels. The secondary surveillance radar (SSR) Mode S, which is increasingly being used for surveillance in high-density airspace, has the capability of transmitting digital data between air and ground. An aeronautical telecommunications network (ATN) will provide for the interchange of digital data between end users over dissimilar air-ground and ground-ground communication subnetworks. The regular use of data transmission for ATM purposes will introduce many changes in the way that communication between air and ground takes place, and at the same time offer many new possibilities and opportunities.

Developing the Future Aviation System

Navigation

Improvements in navigation include the progressive introduction of area navigation (RNAV) capabilities along with the global navigation satellite system (GNSS). These systems provide for world-wide navigational coverage and are being used for world-wide enroute navigation and for certain types of less critical approaches, known as non-precision. With appropriate augmentation systems and related procedures, these systems will eventually support most precision approaches to airports.

Surveillance

Traditional SSR modes will continue to be used, along with the gradual introduction of Mode S in both terminal areas and high-density continental airspace. The major breakthrough however, is with the implementation of automatic dependent surveillance (ADS). ADS allows aircraft to automatically transmit their position, and other data, such as heading, speed and other useful information contained in the flight management system (FMS), via satellite or other communication links, to an ATC unit where the position of the aircraft is displayed somewhat like on a radar display. ADS can also be seen as an application that represents the true merging of communications and navigation technologies, and, along with ground system automation enhancements, will allow for the introduction of significant improvements for ATM, especially in oceanic airspace. Software is currently being developed that would allow this data to be used directly by ground computers to detect and resolve conflicts between aircraft. Eventually, this could lead to clearances being negotiated between airborne and ground-based computers with little or no human intervention.

ADS-broadcast (ADS-B) is another concept for dissemination of aircraft position information. Using this method, aircraft periodically broadcast their position to other aircraft as well as to ground systems. Any user, whether airborne or on the ground, within range of the broadcast, receives and processes the information. All users of the system have real-time access to precisely the same data, via similar displays, allowing a vast improvement in traffic situational awareness.

Air traffic management

In considering implementation of new communications, navigation and surveillance systems and all of the expected improvements, it can be seen that the over-all main beneficiary is likely to be ATM. More appropriately, the advancements in CNS technologies will serve to support ATM. When referring to ATM in the future concept, much more than just ATC is meant.

In fact, ATM refers to a system's concept of management on a much broader scale, which includes ATS, air traffic flow management (ATFM), airspace management (ASM) and the ATM-related aspects of flight operations.

An integrated global ATM system will fully exploit the introduction of new CNS technologies through international harmonization of standards and procedures. Ultimately, this would enable the aircraft operators to conduct their flights in accordance with their preferred trajectories, dynamically adjusted, in the most optimum and cost-efficient manner.

Future ATM system design

The primary goal of an integrated ATM system is to enable aircraft operators to meet their planned times of departure and arrival and adhere to their preferred flight profiles with minimum constraints and no compromise to safety.

The planning for implementation of CNS technologies is well under way in varying degrees around the world. ICAO is focusing the transition toward a clear concept of how to integrate the CNS elements into a coherent and seamless global ATM system. One of the guiding principles for transition and implementation of CNS/ATM systems is that safety levels must be improved. However, while safety is a primary objective, it cannot be considered in isolation of the need to provide for an orderly and efficient flow of air traffic. ATM systems will be developed and organised to overcome the shortcomings listed in this paper and to accommodate future growth, so as to offer the best possible service to all airspace users and to provide adequate economic benefits to the civil aviation community.

Airport surface movement area

Increased airport capacity is a major objective of the future ATM system. The design of the system will contribute to this goal by implementation of techniques, procedures and technologies that fully utilize scarce capacity resources allowing a higher throughput of traffic and maximizing both approach and departure operating efficiencies. Sophisticated automation and an air-ground digital data link will be required to make maximum use of capacity and to meet throughput requirements by improving the identification and predicted movement of all vehicles on the airport movement area, to include conflict advisories. Additionally, increasing levels of collaboration and information sharing between users and ATM providers will create a more realistic picture of airport departure and arrival demand, allowing users to make scheduling and flight planning decisions.

Advanced surface movement guidance and control systems (A-SMGCS) will be used for routing, guidance, surveillance and control of aircraft and

Developing the Future Aviation System

vehicles in order to maintain acceptable movement rates under all weather conditions, while improving the required level of safety. A-SMGCS will also help to ensure that departing aircraft arrive at the holding point of their assigned runway in time to meet departure times required for ATFM.

A-SMGCS will give ATM providers an enhanced surveillance capability of the aerodrome surface and will assist in taxi route planning and conflict detection/resolution. Surface movement management will become automated with aircraft/vehicle positional information being derived from on-board systems such as ADS-B. In this way, situational displays for the cockpit and vehicles will be updated by global navigation satellite systems (GNSS) and provide pilots and vehicle drivers with more precise ground manoeuvring guidance in low visibility and/or high traffic density conditions.

Terminal control operations during climb and descent

The use of curved approaches may eliminate some of the constraints imposed by centre line approach procedures dictated by the limitations of ILSs. In certain situations, this will be essential to eliminate conflicts among approach operations involving adjacent airports and will allow each airport to operate independently. These techniques will also lead to added flexibility in controlling the noise footprint of airport traffic operations. Independent instrument flight rules (IFR) approaches to parallel runways spaced as closely as 760 m (2 500 ft) or less might be routinely based on high data rate SSR and other surveillance techniques and improved monitor controller displays. This will provide capacity increases in instrument meteorological conditions (IMC) at locations with such runway configuration. In addition, many communities will take advantage of this new capability by constructing closely spaced parallel runways that conserve land. Automation tools will assist air traffic managers in establishing efficient flows of approaching aircraft for parallel and converging runway configurations.

Improved metering, sequencing and spacing of arrival traffic using automated metering devices, will increase runway capacities in IMC to a level approaching present runway capacities in visual meteorological conditions (VMC).

Enroute Operations

The flow management process will monitor capacity resources and demand at airports and in terminal and en-route airspace and will implement flow management strategies, where required, to assure that excessive levels of congestion do not develop. The tactical management process will monitor aircraft movements to assure conformance with flight plans and to identify and resolve problems such as imminent separation violations and aircraft incursions into special use airspace. Clearances involving position and time and the ATM data link interface with flight management computers will be principal tools in assuring that ATM constraints are met with minimum deviation from user-preferred trajectories. An increased ability to accommodate user-preferred flight profiles and schedules will be gained through improved decision support tools for conflict detection and resolution and for flow management. Terminal and enroute ATM functions will be integrated to provide a system in which traffic flows smoothly into and out of terminal areas.

Automated and seamless co-ordination supported by ATS inter-facility data communications (AIDC) will present a transparent system to users. In addition, data link will also be used to transmit weather observations from appropriately equipped aircraft and to provide a variety of aviation information to the cockpit including weather information and information on the status of facilities and airports. Departure and arrival route structures will be expanded to permit greater use of RNAV departure and arrival routes.

Oceanic operations

Oceanic operations provide a full breadth of opportunity to benefit from new technologies allowing these operations to experience significant improvements. The overall goal will be to make oceanic ATM operations as flexible as reasonably possible in accommodating user preferred trajectories.

Future oceanic ATM operations will make extensive use of ADS, HF and satellite-based digital communications, GNSS, aviation weather system improvements and collaborative decision making techniques. These new capabilities will permit flexible routing and dynamic modifications to aircraft routes in response to changes in weather and traffic conditions. RNAV, based on GNSS will allow increased capacity through reduced separation minima in the longitudinal and lateral axes. Reduced vertical separation minimum (RVSM) above FL 290 will also increase capacity. More precise monitoring of aircraft, including various conformance monitoring techniques, will allow separation assurance to be accomplished with the aid of decision support systems and visual display systems.

Emerging concepts

In order to attain the goals of future ATM, which is a seamless, global ATM system, progress has to be made on several fronts. Furthermore, emerging concepts such as situational awareness, separation assurance and collision

Developing the Future Aviation System

avoidance will also have to be defined and agreed to globally. The work is being progressed by various groups and will be brought together under the ICAO umbrella where world-wide standards for safety, efficiency and regularity can be agreed to.

Free flight

One such effort toward implementation of a concept aimed at added levels of autonomous flight is being developed in the United States and is known as 'free flight'. The RTCA Task Force 3 on Free Flight Implementation, the group responsible for creation of the concept, bases its work on the identified CNS/ATM systems technologies. RTCA defines free flight as follows (RTCA, 1995):

> .. a safe and efficient flight operating capability under instrument flight rules (IFR) in which the operators have the freedom to select their path and speed in real time. Air traffic restrictions are only imposed to ensure separation, to preclude exceeding airport capacity, to prevent unauthorized flight through special use airspace (SUA), and to ensure safety of flight. Restrictions are limited in extent and duration to correct the identified problem. Any activity which removes restrictions represents a move toward free flight. This suggests that each user is granted both maximum flexibility and guaranteed safe separation. the goal is not only to optimize the system but to open the system for each user to self-optimize. Self-optimization is the key to understanding the extent of free flight's reach as well as free flight's challenges. Free flight is not limited to airspace - its spatial constraints are chock-to-chock, but free flight reaches into a flight's pre-history by providing increased flexibility in flight planning.

In the United States, it is believed that free flight can provide the needed flexibility and capacity for the foreseeable future (approximately the next 50 years). At its basis, the concept enables optimum (dynamic) flight paths for all airspace users through CNS/ATM technologies and the establishment of ATM procedures that maximize flexibility while assuring positive separation of aircraft.

The primary difference between today's direct route clearance and free flight will be the pilot's ability to operate the flight without specific route, speed, or altitude clearances. Restricting the flexibility of the pilot will only be necessary when: potential manoeuvres may interfere with other aircraft operations/special use airspace; traffic density at busy airports or in congested airspace precludes free flight operations; unauthorised entry of a special use airspace is imminent, or, safety of flight restrictions are considered necessary by the air traffic controllers.

In the free flight system, a flight plan will be available to the air traffic service provider to assist in flow management, but will no longer be the basis for separation. It is possible, and highly desirable, to shift from a concept of strategic (flight path based) separation to one of tactical (position and velocity vector based) separation. There even may be instances included in the system's design where separation assurance shifts to the cockpit. When this occurs, there will be clearly defined responsibility (pilot or controller) for traffic separation.

In the free flight concept each aircraft flies a dynamic, optimum flight path; making full use of on-board systems. Position and short-term intent information is provided to the air traffic service provider who performs separation monitoring and prediction functions. The air traffic service provider intervenes to resolve any detected conflicts. Short-term restrictions are used only when two or more aircraft are in contention for the same airspace. In normal situations, aircraft manoeuvring is unrestricted. Separation assurance may be enhanced by appropriate on-board systems. In the future system, each aircraft will be surrounded by two zones.

The smaller zone, designated the protected zone, must remain sterile to assure separation. The size of this smaller zone is a direct reflection of position determination accuracy. The outer zone, designated the alert zone, is used to indicate a condition where intervention may be necessary. The size of the alert zone is determined by aircraft speed and performance and by CNS/ATM capabilities.

An aircraft separated from other aircraft, so that its alert zone is clear, is free to change course, altitude, or speed at will. After any change, a revised plan will be data linked to the ground system for planning purposes. When alert zones of two or more aircraft touch, the air traffic management system will assess the potential for conflict and issue preventive advisories or resolution instructions as necessary. Highly accurate aircraft position and velocity vector information and advanced automation allow the shift to near-term conflict identification and resolution.

ATM strategy for 2000+

The European Organization for the Safety of Air Navigation (EUROCONTROL) has defined the future European ATM System (EATMS) at the request of the European Civil Aviation Conference (ECAC) Transport Ministers. This is also in follow-up of, and in accordance with, the CNS/ATM systems as developed within ICAO (EUROCONTROL, 1997a & 1997b).

The ATM Strategy for 2000+ provides an effective framework of top-down change within which National plans can be developed. The focus is on the progressive introduction of a number of operational improvements which keep pace with the traffic increase while providing early benefits for the airspace users whenever possible.

It is essential to define clear strategic targets and objectives which the new ATM system must meet to satisfy air transport needs.

To overcome the shortfalls of the present ATM organization and concepts and provide an effective platform for meeting the performance targets means having to re-visit the ways in which ATM functions and the scope of the services it provides. The EATMS target operational concept towards which the ATM 2000+ Strategy is aimed, focuses on providing extra capacity and improving ATM services by:

- managing flights from gate-to-gate and allowing the airspace users to operate as closely as possible to their preferences;

- enhancing flexibility and efficiency by optimising the trajectory of the flight;

- improving collaborative decision making by a permanent dialogue between ATM, Aircraft Operations Centres (AOCs), pilots and Airport Operations;

- adapting the available capacity to meet demand through more flexible use of airspace, improved planning, extensive use of enhanced computer tools and automation of certain ATM processes, and new or revised procedures governing the roles and allocation of responsibilities for separation and traffic sequencing.

Progressive improvement in these areas will also contribute to enhanced safety, extend the principles of uniformity and seamless services, and help reduce aviation related environmental pollution.

The target operational concept is predicated on layered planning, based around a strategically derived daily airspace plan, and collaborative decision making between the involved parties with an evolving change to managing resources rather than demand. ECAC airspace will be regarded as a continuum for airspace planning purposes to optimize the available resources. Airspace divisions will be based on ATM needs rather than on national boundaries, but without compromising sovereignty. Route structures will be used to generate additional capacity in high traffic density areas, but the concept incorporates free flight airspace and user-preferred routings for suitably equipped aircraft whenever capacity considerations allow. There will also be fundamental changes to current roles, both in the air and on the ground; a distribution of responsibilities for separation assurance between the air and ground ATM elements according to aircraft capabilities and the services provided; greater use of computer support tools to cope with increased levels of service and keep ATC and cockpit workload within acceptable levels; and a more dynamic and flexible management of airspace.

Benefits of the future Air Traffic Management System

A large number of technologically-related opportunities and benefits are now available for implementing a world-wide ATM system that will improve ATM services to better meet user requirements. The new technologies and associated ATM procedures will also provide for an improvement in controller productivity and over-all enhancement of the work environment.

CNS/ATM systems will improve the handling and transfer of information, extend surveillance using ADS and improve navigational accuracy. This will lead, among other things, to reductions in separation between aircraft, allowing for an increase in airspace capacity.

Advanced CNS/ATM systems will also see the implementation of ground-based computerized systems to support increases in traffic. These ground-based systems will exchange data directly with flight management systems (FMSs) aboard aircraft through data link. This will benefit the ATM provider and airspace user by enabling improved conflict detection and resolution through intelligent processing and providing for the automatic generation and transmission of conflict free clearances as well as offering the means to adapt quickly to changing traffic requirements. As a result, the ATM system will be better able to accommodate an aircraft's preferred flight profile and help aircraft operators to achieve reduced flight operating costs and delays.

Benefits for the airlines

Benefits of CNS/ATM systems will come to providers and users of the air navigation system through formation of a more close-knit relationship allowing rapid and reliable transmission between ground and airborne system elements. More accurate and reliable navigation systems will also allow aircraft to navigate in all types of airspace and operate more closely together. In anticipation of advantages of CNS/ATM systems, the airlines expect reduced separation standards over oceanic airspace; increased access to remote areas; the gradual introduction of 1 000 ft. vertical separation above 29 000 ft.; increased opportunities for more dynamic and direct routings; and an over-all enhancement of safety.

Developing the Future Aviation System

Benefits for the states that provide the air navigation infrastructure

For those States that provide and maintain extensive ground infrastructures, a reduction in the over-all cost of operation and maintenance of facilities is expected, as the traditional ground systems become obsolete and satellite technology is increasingly employed. They will also benefit from enhanced safety. CNS/ATM provides a timely opportunity for developing States to enhance their infrastructure to handle additional traffic with minimal investment. Many of these States have large areas of available but unusable airspace, mainly because of the expense involved in purchasing, operating and maintaining the necessary ground infrastructures. CNS/ATM systems will afford them opportunities to modernise inexpensively, which includes the provision of precision and non-precision approaches.

Benefits to the General Aviation Community

General aviation and utility aircraft will find increasing access to avionics equipment that will allow them to operate in flight conditions, as well as into and out of airports, that they would normally have been prohibited from using because of the operating cost and associated requirements.

Furthermore, as a result of implementing CNS/ATM systems, many remote areas that are currently inaccessible to most general aviation aircraft because of their inability to communicate or safely navigate over them, would become accessible.

Indirect benefits

In addition to the direct benefits listed above, there are also many indirect benefits such as:

- lower fares and rates;

- passenger time savings;

- environmental benefits;

- transfer of high technology skills;

- productivity improvements and industry restructuring;

- stimulation of related industries;

- enhanced trade opportunities; and

- increased employment.

References

European Organisation for the Safety of Air Navigation, (1997a), *EATMS Operational Concept Document (OCD)* (Eurocontrol Doc. No.xx, Issue 1.0), Eurocontrol, Brussels, Belgium.

European Organisation for the Safety of Air Navigation, (1997b), *ATM Strategy for 2000+* (Eurocontrol Doc. No.xx, Issue 1.0), Eurocontrol, Brussels, Belgium.

International Air Transport Association, (1995), *Concept for Air Traffic Management in the Future Air Navigation Systems* (Doc. GEN/3140), International Air Transport Association, Montreal, Canada.

International Civil Aviation Organization, (1991a), *Tenth Air Navigation Conference (1991)* (Doc 9583), International Civil Aviation Organization, Montreal.

International Civil Aviation Organization, (1993a), *Report of the Fourth Meeting of the Special Committee for the Monitoring and Co-ordination of Development and Transition Planning for the Future Air Navigation System (FANS Phase II)* (Doc 9623), International Civil Aviation Organization, Montreal.

International Civil Aviation Organization, (1998), *The Global Air Navigation Plan for CNS/ATM Systems*, International Civil Aviation Organization, Montreal.

RTCA, (1995), *Report of the Board of Directors' Select Committee on Free Flight*, RTCA Incorporated, Washington, D.C.

The author

Vince Galotti has had a distinguished career from being an air traffic controller, graduate of Embry-Riddle Aeronautical University to his present position as a Technical Officer in the Air Navigation Bureau of ICAO. Having many publications in the Air Traffic field his recent book about FANS was recently issued by the Ashgate Publishing Group.

11 Improving capacity – implementation of the FANS CNS/ATM system in the Asia/Pacific region

Brian O'Keeffe

Introduction

The FANS CNS/ATM system has been eagerly embraced by the air traffic service providers and airlines in the Asia/Pacific Region as the only viable means of overcoming the problems of the past and catering for increasing air traffic. It has always been difficult and often impossible to provide high quality communications, navigation and surveillance services throughout the Region from the old CNS because of the limitations imposed by oceans, mountains, jungles, deserts, etc., coupled with a lack of resources and expertise. The result has been a less than optimum air traffic management system. To compound this, the Region has the highest predicted growth rate for air traffic in the world where the IATA estimates show that, by the year 2010, over one half of the world's international passenger traffic will be in the Asia/Pacific Region.

It needs to be emphasised that the Asia/Pacific Region is looking for the FANS CNS to provide the much needed communication, navigation and surveillance services over the larger areas where little currently exists and probably never will, if the old CNS were to continue. The Region thus differs markedly from North America and Europe where there is a high quality ground based infrastructure in place based on the old CNS. Thus the Asia/Pacific Region is not looking for a replacement for the old CNS, but rather for the FANS CNS to provide services which cannot be provided today. It is therefore not surprising that the air traffic service providers and airlines have recognised the essentially of the FANS CNS and have pressed ahead with its implementation to harvest early benefits.

The FANS Committee (Phase 1)

In the early 1980's, the aviation community came to recognise the limitations of the then current CNS systems which were causing problems and would act as a constraint on the growth of civil aviation in the coming years. This was not surprising because the existing system which had served aviation well in the past decades was made up of elements which, with a few exceptions, had been in use in the 1940's. They were more than 40 years old, and the existing system was indeed showing signs of strain.

In response to the need to overcome the problem, ICAO established in 1983 the Special Committee on Future Air Navigation Systems to develop a solution which would :

- overcome the present problems,

- cater for future growth,

- take aviation into the 21st century.

The first meeting of the FANS Committee took place at ICAO's Headquarters in Montreal in July 1984. Mr Jan Smit from the Kingdom of the Netherlands was elected Chairman of that meeting and three subsequent meetings of the FANS Committee over its first four years of operation. Mr Smit's great sense of purpose acted as the necessary catalyst that lead for the Committee to complete the system concept on four years.

One of the early tasks of the FANS Committee in its first phase was to investigate the existing CNS system to find the basic causes of the problems. The Committee concluded that the short comings of the existing system were due to :

- the propagation limitations of current line-of-sight systems and the variability of propagation characteristics of other systems,

- difficulties in implementing present CNS systems in large parts of the world,

- limitations of voice communications and lack of air-ground data interchange systems.

The FANS Committee concluded that, to overcome present and future problems, new concepts and new systems were needed. It was not surprising that the system designed by the FANS Committee was based largely on

Developing the Future Aviation System

satellite technology which had matured in the 1980's. However, it was also recognised that some of the best of the terrestrial systems, such as VHF and SSR, would continue to provide a useful service in terminal areas and other busy airspaces. The FANS CNS is therefore a mixture of satellite and ground based systems to achieve an overall optimum result. It may be surprising to some that the FANS CNS is not a single system, but is made up of a variety of elements. It is thus very flexible in its application, because it can be implemented using any or all, or nearly any combination of elements from the communications, navigation and surveillance groups. The FANS CNS can be thought of as a menu of elements as follows :

Communications : Satellite ; VHF ; SSR Mode S.

Navigation (to meet the required navigation performance) : GNSS ; IRS/INS;   ILS/MLS ; barometric altimetry.

Surveillance : ADS ; SSR.

Countries with large and busy airspaces will probably implement most of the above elements. However, countries with small continental airspaces could implement the complete FANS CNS by using VHF (with data link) for communications and surveillance and by using GNSS for navigation. Thus the implementation can be tailored to a countries and aircraft operators requirements. While this makes the FANS CNS very flexible and cost effective, it can cause bewilderment to some countries and airlines because of the large choice of elements. Fortunately, such problems are largely overcome by countries and aircraft operators participating in Regional Planning to make decisions of benefit to all.

Benefits of enhanced ATM

The benefits to air traffic service providers and airlines will be achieved directly from the implementation of the FANS CNS, but even greater benefits will come from the improvements in ATM which are made possible by the FANS CNS. The benefits include :

- maintaining and improving safety with traffic growth,

- provision of CNS in a more cost-effective manner,

- global provision of CNS in a more uniform manner,

- the improvement of ATM to make more efficient use of airspace and airport capacity and to permit optimisation of flight profiles.

Satellite communication provides direct pilot to controller communication on a global scale which greatly expedites clearances, requests for changes en route, etc. Data link allows large volumes of data to be exchanged between air traffic control and aircraft to allow dynamic replanning of flights for the dynamic optimisation of flight profiles. The extension of surveillance made possible by ADS contributes to the reduction in separations and to the optimisation of flight profiles. CNS and ATM are closely coupled and are thus known as CNS/ATM. All these are of great importance in the Asia/Pacific Region.

The FANS Phase II committee

The first FANS Committee, having completed the development of the new CNS system concept in May 1988, recommended that ICAO establish a new Committee:

> to advise on the overall monitoring, coordination of development and transitional planning to ensure that the implementation of the future CNS system takes place on a global basis in a cost effective manner and in a balanced way between air navigation systems and geographical areas.

In 1989 ICAO established the FANS (Phase II) Committee with the tasks of:-

- identifying and recommending acceptable institutional arrangements,

- developing a global co-ordinated plan,

- monitoring research and development programs, trials and demonstrations.

I had the honour of being elected Chairman of the meetings of the FANS (Phase II Committee) as well as the FANS Interim Committee which operated in 1988 and 1989 between the two FANS Committees so as not to lose the momentum of the development. Five Rapporteurs were appointed to progress the FANS Working Groups: Ron North (Canada), Norm Solat (USA), George Paulson (UK), Sven Andresen (Denmark) and Olivier Carel

Developing the Future Aviation System

(France). The success of the FANS (Phase II) Committee was in no small part due to the tireless efforts of the Rapporteurs.

Over the nearly ten years of operation of the FANS Committees, the membership comprised representatives of some 40 countries and international organisations in what became an international collaborative effort on a global scale. The great strengths of the FANS Committees were the high level of representatives, the multidisciplinary composition of the Committees, and the tremendous resources which its members, observers and advisors brought to the committees for the tasks.

In 1991, ICAO convened a world-wide meeting, The Tenth Air Navigation Conference, which endorsed the FANS concept.

Planning for implementation

The transition to FANS CNS/ATM is a large undertaking which fundamentally alters how air traffic control and aircraft do business. The elements in the transition will be:

- from *GROUND BASED*   to   *SATELLITE BASED*

- from *INDIVIDUAL*   to   *GLOBAL*

- from *ANALOG (SPEECH)*   to   *DIGITAL (DATA)*

As the FANS CNS/ATM is fundamentally a global system, it was necessary to begin planning on a global basis with a broad global plan. This was completed by the FANS Phase II Committee in late 1993 as the "Global Coordinated Plan for the Transition to the ICAO CNS/ATM Systems" and consists of:

- a description of the FANS system and benefits,

- transition,

- key events necessary for orderly global implementation and co-ordination information,

- information on institutional arrangements, research and development activities and ATM consideration.

Improving capacity

The actual implementation of FANS CNS/ATM is done by national civil aviation bodies (either on their own or as a group) and the airlines. These are the organisations which currently provide and operate most of the ground and airborne parts of the system. Of course, they cannot do this in isolation because co-ordination is needed, particularly between adjacent countries and airlines which fly through the countries airspace. If co-ordination is lacking, there will be different time scales and systems, resulting in discontinuities at the airspace boundaries which would negate many of the benefits which the new system offers. So while the major activity on FANS CNS/ATM began at the global level, the focus has now shifted to the regional level where the national aviation bodies, airlines, service providers, etc. come together to co-ordinate their plans. Implementation planning is thus at three levels :

- The Global Coordinated Plan,

- Regional Plans,

- States/Airlines Plans.

The Asia/Pacific Implementation Regional Plan

The Asia/Pacific Region contains great diversity and stretches from Pakistan to the United States and from Mongolia to the South Pole. However, the Region was quick to recognise the benefits of the FANS CNS/ATM and were united in pressing for its implementation. Thus when the Asia/Pacific Air Navigation Planning and Implementation Regional Group (APANPIRG) was established by ICAO in 1992, it immediately set up the CNS/ATM Sub group to produce the CNS/ATM plan for all the countries in the Region. The broad plan was produced in 1994 and drew heavily on the FANS Committee's Global Coordinated Plan. The Regional plan contains traffic forecasts, the description of the FANS CNS/ATM, transition guidelines, trials and developments and an implementation summary. The plan is updated annually to keep pace with progress of developments.

In 1994, the CNS/ATM Sub group was reformed to be the "CNS/ATM Implementation Coordination Sub group" to emphasise that implementation had begun and was being co-ordinated on a Regional basis. For this phase the Sub group developed a new planning methodology which is essentially a "top down" approach where the air traffic service providers and the airlines meet to decide the air traffic system requirements and the time scales for implementation. This is a great improvement on the old methodology for regional planning where the regional plan was assembled by putting together

Developing the Future Aviation System

the plans of each country. The result was inevitably a patchwork, often with missing links. The FANS CNS/ATM and the new methodology has given the Region the tools and opportunity to plan on a larger scale to better meet the needs of the air traffic service providers and the airlines. For example, the optimisation of the 14 hour flight from Los Angeles to Sydney needs to be based on the whole route from origin to destination and *not* by independent segments through each countries' airspace.

The sub-group therefore divided up the Asia/Pacific Region into nine geographical areas based on the major traffic flows. These are shown in Table 11.1, together with the flight information regions included in each geographical area. This then allows planning for the ground and airborne systems needed from origin to destination in each of the nine geographical areas. The Sub-group then identified 26 operational enhancements in the air traffic management system. For each of these, the corresponding ground and airborne systems needed to achieve that enhancement over the whole traffic flow were then determined. The resulting generic table is shown in Table 11.3 and the explanation for the terms used are in Table 11.2. There will, of course, be differences in the systems required to achieve the ATM enhancements in each of the nine geographic areas because of the traffic volumes and mix, route complexity, and implementation time frames. The Sub-group has therefore developed a set of tables, based on the master generic table, for each of the nine geographical areas. The table for the traffic flows between Australia/New Zealand and North America (via South Pacific) is shown in Table 11.4 and Asia - North America (via the North Pacific and the Russian Far East) is shown in Table 11.5.

Of particular interest in the Tables is they show what has been achieved already together with near and medium term dates for implementation for further enhancements. Thus it will be seen in Table 11.4 that :

- Flexible tracks on the Los Angeles - Sydney/Auckland routes have already been implemented with resulting savings to the airlines.

- Dynamic re-routing is targeted for implementation during 1997/98.

- Dynamic User Preferred Routes are targeted for implementation in 1998-99 and this is expected to achieve a large part of the benefits of "Free Flight" for the airlines.

- 50 NM longitudinal and lateral separation are targeted for implementation during 1997.

Improving capacity

Some recent milestones

A major milestone for the early implementation of the FANS CNS/ATM was achieved in mid 1995 with the FAA certification of the FANS 1 package for the Boeing 747-400 which includes :

- multi sensor navigation using IRS and GPS,

- ADS,

- Direct pilot-controller data communications (by VHF and satellite).

The certification trials were done jointly in the Asia/Pacific Region by :

- Boeing as the aircraft manufacturer,

- Honeywell as the avionics provider,

- Qantas as the provider of the B747-400 for the flight trials,

- INMARSAT, ARINC and SITA as the communication service providers,

- Airservices Australia as the provider of the ground ADS and data link systems to receive, process, record and transmit data messages with the Qantas aircraft.

Some 300 FANS1 packages have now been ordered for the Boeing 747-400 and over 150 already fitted. The FANS 1 package is available for the B777 and will be available for the Boeing 767 and 757 by the end of 1997. Similar certifications have been done by the JAA in Europe and similar airborne systems are available for Airbus aircraft, particularly the Airbus 340 and 330. IATA estimates there will be some 600 aircraft fitted with FANS 1 (or equivalent) by the end of 1998. On the ground, direct pilot-controller voice and data communications are already being used as the primary means of ATC from Oakland, Tahiti, Auckland, Nadi and Brisbane ATC centres.

Another successful achievement has been the implementation of satellite navigation in Fiji. Some years ago, the Civil Aviation Authority of Fiji recognised that the only cost effective way to obtain en route navigation and approaches under instrument conditions for domestic operations was to use GPS. As a result of the GPS implementation that has taken place in Fiji, they have had instrument procedures for en-route, cloudbreak, terminal arrival, GPS step descents and GPS departures *since 1993*. There are also 'GPS only' routes. Nearly all the Fiji registered aircraft are fitted with GPS as a primary means of navigation and this may be made mandatory. The overall result has been a considerable improvement in the reliability and efficiency of aircraft operations in Fijian airspace.

A beneficial outcome already achieved through the use of flexible tracks has been the reduction in aircraft having to make technical stops for fuel for flights across the Pacific. Much greater benefits can be expected when dynamic rerouting is introduced in 1997 and the dynamic user preferred routes in 1998-99.

Future implementation

The question may be asked as to whether there will be sufficient FANS CNS/ATM systems available in the Asia/Pacific Region to proceed with confidence with implementation to achieve the operational enhancements set out in Table 11.3. The answer is certainly - YES !

For communication satellites, we already have the INMARSAT Pacific and Indian Ocean satellites with the latest generation of these satellites being brought into service. Japan has a firm program for the provision of satellites for aeronautical purposes - MTSAT to come into service at the turn of the century. Other countries in the Region are considering launching additional satellites which would provide for aeronautical communications.

For navigation, GPS and GLONASS are already available. Airborne integrity augmentation systems are already certificated. Ground based augmentation systems are being built. Additional GPS navigation signals will be available from geostationary satellites already launched or planned, e.g. INMARSAT and MTSAT. The additional sources of GPS navigation information available from these four satellites will greatly enhance airborne augmentation systems. Thus the Region will have access to a more than adequate number of satellites for both communication and navigation.

On the ground, VHF data link stations are expanding rapidly, particularly in the Russian Far East and China. Data link ground systems are already in operation across the Pacific and more are being implemented in Singapore, Malaysia, Hong Kong, China, Japan, India and elsewhere. ADS is being added to the data link systems already installed to meet the requirements of reduced separation and dynamic user preferred routes.

Airborne equipments (the FANS 1 package and its equivalent) are already available for the larger aircarrier aircraft and programs are underway to extend this to other aircraft. Several equipments are under development for both the older and smaller aircraft. So there are sufficient space, airborne and ground systems available to proceed with the ATM enhancements to obtain benefits.

It is, of course, essential that there is the continuing commitment from the air traffic service providers and the airlines to implement and operate the FANS CNS/ATM in accordance with the Regional Implementation Plan. If not, there will be "missing links" on major routes and discontinuities at airspace boundaries resulting in less than expected benefits. It has been recognised by the CNS/ATM Implementation Coordination Sub-group that some countries with limited expertise and resources will need considerable help for their implementation of CNS/ATM. Sub-group members with available expertise and resources have found ways to assist on the basis of a mutual help system where those with expertise and resources can help those who have not.

There is also a need for interregional co-ordination because major traffic flows often pass over more than one region. Some of these routes are very long and thus the optimisation of the flight profile has large economic benefits. Fortunately, this has been recognised by ICAO which has established a body called ALLPIRG comprised of the chairpersons and secretaries of all the Regional Planning groups, together with representatives of the aviation industry. ALLPIRG had its first meeting in April 1997 to develop the machinery for interregional co-ordination.

Conclusion

The FANS CNS/ATM is vitally important for the Asia/Pacific Region, where the lack of wide coverage, high quality conventional CNS has resulted in a less than optimum ATM. Implementation of the FANS CNS/ATM is the only viable means of overcoming present problems and catering for the large predicted growth in air traffic.

Fortunately, the Asia/Pacific Region is well advanced in the implementation of the FANS CNS/ATM. There is a region wide plan with agreed implementation dates for ATM enhancements in the short and medium term to effect the needed co-ordination between the air traffic service providers and airlines. Further details are being developed and the plan is updated on a yearly basis. Assistance is being given to ensure implementation takes place over the whole of a route. Most importantly, implementation is well underway, the operational use of FANS CNS/ATM is taking place and benefits are being obtained.

Table 11.1
Geographical areas based on major traffic flows

ASIA/AUSTRALIA – AFRICA

This geographical area encompasses the major traffic flows between Africa, Asia and Australia.

It includes the following FIR/UIRs: Melbourne, Jakarta, Singapore, Kuala Lumpur, Bangkok, Yangon, Madras, Colombo, Male, Bombay, [and African FIR/UIRs].

AUSTRALIA/NEW ZEALAND – ASIA

This geographical area encompasses the major traffic flows from Australia and New Zealand to Asia and Indonesia north to Tokyo.

It includes the following FIR/UIRs: Nadi, Auckland, Nauru, Honiara, Oakland, Brisbane, Port Moresby, Melbourne, Biak, Ujung Pandang, Bali, Jakarta, Singapore, Kota Kinabalu, Manila, Ho Chi Minh, Hanoi, Phnom-Penh, Vientiane, Bangkok, Kuala Lumpur, Yangon, Hong Kong, Taipei, Naha, Tokyo, Shanghai, Taegu, Guangzhou, Wuhan, Beijing.

ASIA - EUROPE (North of Himalayas)

This geographical area encompasses the major traffic flows between Asia and Europe via north of the Himalayas.

It includes the following FIR/UIRs: Bangkok, Ho Chi Minh, Phnom-Penh, Hanoi, Vientiane, Yangon, Kathmandu, Guangzhou, Kunming, Wuhan, Beijing, Urumqi, Shanghai, Shenyang, Lanzhou, Hong Kong, Taipei, Naha, Tokyo, Taegu, Pyongyang, Ulaanbaatar, Almaty, [Russian Federation FIRs, and European FIRs].

ASIA - EUROPE (South of Himalayas)

This geographical area encompasses the major traffic flows between Asia and Europe via south of the Himalayas. It includes the following FIR/UIRs: Manila, Ho Chi Minh, Hanoi, Vientiane, Phnom-Penh, Bangkok, Yangon, Ujung Pandang, Bali, Kota Kinabalu, Jakarta, Singapore, Kuala Lumpur,

Hong Kong, Colombo, Madras, Calcutta, Dhaka, Kathmandu, Kunming, Delhi, Bombay, Lahore, Karachi, [and Middle East/European FIR/UIRs].

ASIA - NORTH AMERICA (via North Pacific and the Russian Far East) – (see Table 11.4).

This geographical area encompasses the major traffic flows from North America to Asia via the North Pacific air routes and the Russian Far East.

It includes the following FIR/UIRs: Anchorage, Canadian FIR/UIRs, Russian Far East (Anadyr, Mys Schmidta, Okha, Tilichiki, Petropavlovsk-Kamchatsky, Khabarovsk, Vladivostok, Yuzhno-Sakhalinsk, Magadan), Pyongyang, Taegu, Shenyang, Beijing, Hong Kong, Guangzhou, Wuhan, Shanghai, Ulaanbaatar, Tokyo.

ASIA - NORTH AMERICA (via Central Pacific)

This geographical area encompasses the major traffic flows between the United States (including Honolulu) and Canada to North and East Asia.

It includes the following FIR/UIRs: Oakland (North of 21N to south of NOPAC Routes), Vancouver, Tokyo, Manila, Taipei, Hong Kong, Naha.

AUSTRALIA/NEW ZEALAND - SOUTH AMERICA

This geographical area encompasses the major traffic flows between South America and New Zealand/Australia.

It includes the following FIR/UIRs: Brisbane, Auckland, Nadi, Tahiti, [and South American FIR/UIRs].

AUSTRALIA/NEW ZEALAND - NORTH AMERICA (via South Pacific) – (see Table 11.5)

This geographical area encompasses the major traffic flows between North America, Australia/New Zealand, and the South Pacific Islands.

It includes the following FIR/UIRs: Oakland (southern region), Nadi, Nauru, Honiara, Auckland, Tahiti, Brisbane, and Port Moresby.

Developing the Future Aviation System

SOUTH EAST ASIA - NORTH EAST ASIA

This geographical area encompasses the major traffic flows between South East Asia and China and Japan.

It includes the following FIR/UIRs: Ujung Pandang, Bali, Jakarta, Singapore, Kota Kinabalu, Manila, Ho Chi Minh, Phnom-Penh, Hanoi, Vientiane, Guangzhou, Kunming, Wuhan, Shenyang, Beijing, Bangkok, Kuala Lumpur, Yangon, Hong Kong, Taipei, Naha, Tokyo, Shanghai, Taegu, Pyongyang.

Table 11.2
Explanation of terms used in tables

RNAV ROUTES (code 1a)

ATS Routes in accordance with Annex 11, require aircraft to have an approved Area Navigation System. These are charted routes.

FLEXIBLE TRACKS (code 1b)

RNAV tracks calculated, as required, for minimum flight time. Calculated by one ATS/Airline agency. Same tracks used by all aircraft appropriately equipped. These tracks are not charted.

RNAV PARALLEL ONE WAY ROUTE STRUCTURES (code 1c)

Pairs of RNAV routes which start and end at a common point. Diverge and converge at least 15 degrees to achieve lateral separation minima. Not necessarily referenced to ground navigation aids. These routes are charted.

CNS/ATM ROUTES (code 2)

CNS/ATM routes are seen as routes for which RNP approval and DCPC (voice or data) capability are required. For some routes, ADS capability may also be required. Data link functions supported in airborne and ground systems should be such that end-to-end interoperability and safety are ensured.

FLEXIBLE ROUTES (code 2b)

ATS RNAV routes calculated, as required, for minimum flight time and designated as requiring RNP in accordance with Annex 11. These routes may also require the application of data link and ADS for ATM. These routes are not charted.

DYNAMIC RE-ROUTES (code 2c)

Flexible tracks which can be changed in flight by the use of data link to load a new flight plan into the FMC. Not charted. All aircraft follow the same re-route.

Developing the Future Aviation System

DYNAMIC USER PREFERRED ROUTES (code 2d)

Flexible tracks, which can be changed in flight, which may not be the same for all aircraft flying the same city pair. Not charted. Requires ADS for ATM.

FREE FLIGHT (code 2e)

Free Flight describes a safe and efficient operating capability, in which operators have the freedom to select an optimum flight path in four dimensions. Air traffic restrictions would be imposed only to ensure separation of aircraft, prevent capacity being exceeded, or to otherwise ensure safety.

CNS/ATM APPROVED AIRCRAFT

An aircraft that has the FANS-1 or FANS-A functionality or equivalent (CPDLC per RTCA DO-219, SATCOM, navigation system including GPS certified to an RNP value, ADS per RTCA DO-212, RTA, Autoload capability). Aircraft will also be RVSM approved.

AUTONOMOUS AIRCRAFT

The ultimate aircraft configuration that can communicate, navigate and provide surveillance information anywhere in all ground and airspace configurations.

NOT CURRENTLY BEING EVALUATED

"Not currently being evaluated" means that this ATM operational enhancement has been considered and is recognized as being an enhancement for the geographical area but an evaluation will be carried out at a later date.

NOT BEING CONSIDERED

'Not being considered' means that this ATM operational enhancement will not be considered for evaluation in the short or medium term for various reasons.

Improving capacity

Table 11.3

Generic table for enroute operations

AREA:				Lead Aircraft:	
Current Description:		Communications: Voice: CPDLC:		Note: For each particular operational enhancement, there will be a need for the airlines and the ATS Providers to review existing procedures to identify what new requirements are needed prior to operational implementation	
Separation:		Navigation: Surveillance:			

CODE	ATM OPERATIONAL ENHANCEMENTS	SYSTEM IMPLEMENTATION	REQUIRED FUNCTIONS – AIR	REQUIRED SERVICES - GROUND	NOTES // IMPLEMENTATION DATE
1 RNAV ROUTES					
1 a	RNAV Routes		RNAV Capability		Available today in the Asia Pacific Regions
1 b	Flexible Tracks		FMS or RNAV	Track Generation/Distribution	Available Today in the Asia Pacific Regions
1 c	RNAV Parallel one way Route Structures		RNAV capability		Available Today in the Asia Pacific Regions
2 CNS/ATM ROUTES					
2 a	Fixed Routes		CNS/ATM Approved Aircraft	DCPC (Voice or Data)	Additional requirements may be applicable
2 b	Flexible Routes		CNS /ATM Approved Aircraft	DCPC (Voice or Data)	Additional requirements may be applicable
2 c	Dynamic Re-Route		CNS /ATM Approved Aircraft, with FMS (IRS) CPDLC AOC Data Link Direct Flight Plan Uploads	CPDLC AIR/GND DATA LINK AOC/ACC Data Communication Flight Plan Generation	Utilization dependent on Airspace Complexity
2 d	Dynamic User Preferred Routes		CNS /ATM Approved Aircraft, with FMS (IRS) CPDLC AOC Data Link Direct Flight Plan Uploads	CPDLC AIR/GND DATA LINK AOC/ACC Data Communication Flight Plan Generation ATC Traffic Situation Display.	Utilization dependent on Airspace Complexity.

163

Table 11.3 (continued)

		Autonomous Aircraft	ATM system that is interoperable with Autonomous Aircraft	
2 e	Free Flight			Concept still undergoing definition by ICAO. Utilization dependent on Airspace Complexity.
3 VERTICAL SEPARATION				
3 a	1000' to FL290, 2000' above FL290	ICAO SARPS	ICAO SARPS	Nonstandard application of Cruising levels
3 b	2000' Step Climb above FL290			
3 c	Cruise Climb	FMS (IRS) CPDLC ADS ACAS	CPDLC ADS	Requirements dependent upon airspace complexity. Accomplished today through block altitudes assignments
3 d	1000' Crossing Traffic above FL290	FMS RTA ACAS with Traffic Situation display RVSM Certification	ATC Traffic Situation Display	Requirements dependent upon airspace complexity. For use between RVSM certified aircraft in non RVSM airspace.
3 e	1000' Vertical Separation between FL290 and FL460	RVSM Certification	ICAO SARPS ATC Traffic Situation Display	
4 LONGITUDINAL SEPARATION				
4 a	80nm	RNAV Capability		Available today in the Asia Pacific Regions
4 b	50nm	RNAV or FMS DCPC Voice or Data RNP 10 Approval	DCPC (Voice or Data)	30 Min Reporting Requirement ADS may be required for ATM. Approval process underway for region wide implementation.

Improving capacity

Table 11.3 (continued)

4 c	30nm	Indicative requirements: RNAV or FMS DCPC Voice or Data RNP 4 Approval	DCPC (Voice or Data)	Final requirements TBD 15 Min Reporting Requirement ADS may be required for ATM
4 d	Less than 30nm	Indicative requirements: RNAV or FMS DCPC Voice and Data ☐ RNP Approval ADS	DCPC (Voice and Data) ADS	Final requirements TBD
4 e	10 min	RNAV Capability		Available today in the Asia Pacific Regions.
4 f	7 min	RNAV or FMS DCPC Voice or Data RNP 10 Approval Accurate Time	DCPC (Voice or Data) Accurate Time	Final requirements TBD
4 g	4 min	Indicative requirements: RNAV or FMS DCPC Voice or Data RNP 4 Approval Accurate Time	DCPC (Voice or Data) Accurate Time	Final requirements TBD
5 LATERAL SEPARATION				
5 a	60nm	RNAV Approval		Available today in the Asia Pacific Regions. Navigation performance monitoring may be required.
5 b	50nm	RNAV or FMS RNP 10 Approval		Navigation performance monitoring may be required. Approval process underway for region wide implementation.
5 c	30nm	RNAV or FMS ☐ RNP 4 Approval DCPC voice or data	DCPC (Voice or Data)	Final requirements TBD Navigation performance monitoring may be required.

Table 11.3 (continued)

CODE	ATM OPERATIONAL ENHANCEMENT	SYSTEM IMPLEMENTATION	REQUIRED FUNCTIONS - AIR	REQUIRED SERVICES - GROUND	NOTES:
5 d	Less than 30nm		RNAV or FMS DCPC Voice and Data RNP Approval ADS	DCPC (Voice and Data) ADS	Final requirements TBD Navigation performance monitoring may be required.
6 AIR TRAFFIC MANAGEMENT					Air Traffic Management is a synergy of all three functions outlined below.
6a	Airspace (Dynamic) Management		CNS/ATM Aircraft	Separate data bases for: • Aircraft • AOC • Military reserved airspace • National Security • Environmental • AIS database • Airports • Weather • Traffic • SAR • Rules of the Air	This provides the information that is necessary to create the airspace configuration in which air traffic is managed.
6b	Traffic management (Airspace + aircraft)		CNS/ATM Aircraft	Separate data bases for: • Aircraft • AOC • Military reserved airspace • National Security • Environmental • AIS database • Airports • Weather • Traffic • SAR • Rules of the Air Integrated automation of data base management. Controller/Pilot Interoperability.	Real time management. Purpose is to evolve from unilateral decisions and static flow management, which is based upon a lack of data or data sharing, to an ATS/Pilot/AOC integrated decision that is based on shared real time information. Decisions are based upon achieving optimized gate-to-gate performance to destination airport.

Table 11.3 (continued)

6b ctd.			Automation of traffic movement. Automation of traffic movement optimization. AOC Interface. AIDC.	
6c	Flow Management	CNS/ATM Aircraft	Separate data bases for: • Aircraft • AOC • Military reserved airspace • National Security • Environmental • AIS database • Airports • Weather • Traffic • SAR Integrated automation of data base management. Controller/Pilot interoperability. Automation of traffic movement. Automation of traffic movement optimization. AOC Interface. AIDC	Purpose is to achieve optimization of service balanced with overall air traffic management capacity and safety.

Developing the Future Aviation System

Table 11.4

AREA: AUSTRALIA/NEW ZEALAND - NORTH AMERICA (via South Pacific)

Lead Aircraft: B747-400, B777, B767, B737-700, A340

Current Description:
Generally PANS RAC/DOC 4444
SUPPS/DOC 7030
Separation:
100nm/10min/2000'
Fixed and flexible ATS routes/4000' step climb.

Communications:
Voice: VHF/HF/HF Intermediary /SATVOICE
CPDLC: VHF/SAT
Navigation: FMS/Inertial/GNSS
Surveillance: Procedural/Radar/ADS

Note: For each particular operational enhancement, there will be a need for the airlines and the ATS Providers to review existing procedures to identify what new requirements are needed prior to operational implementation.

CODE	ATM OPERATIONAL ENHANCEMENTS	SYSTEM IMPLEMENTATION	REQUIRED FUNCTIONS - AIR	REQUIRED SERVICES - GROUND	NOTES // TARGET IMPLEMENTATION DATE
1 RNAV ROUTES					
1 a	RNAV Routes	Available today	RNAV Capability		// Implemented
1 b	Flexible Tracks	LAX-SYD-LAX LAX-AKL-LAX	FMS or RNAV	Track Generation/Distribution	Track Generation systems not needed by all States // Implemented
1 c	RNAV Parallel one way Route Structures	SYD-AKL-SYD	RNAV capability		// Implemented
2 CNS/ATM ROUTES					
2 a	Fixed Routes	Not currently being evaluated	CNS/ATM Approved Aircraft	DCPC (Voice or Data)	Additional requirements may be applicable. Data is the intended primary means of communication // Implementation TBD
2 b	Flexible Routes	Not currently being evaluated	CNS /ATM Approved Aircraft	DCPC (Voice or Data)	// Implementation TBD

Improving capacity

Table 11.4 (continued)

2 c	Dynamic Re-Route	LAX-SYD-LAX LAX-AKL-LAX NAN-LAX-NAN NAN-YVR-NAN	CNS /ATM Approved Aircraft, with FMS (IRS) CPDLC AOC Data Link Direct Flight Plan Uploads	CPDLC AIR/GND DATA LINK AOC/ACC Data Communication Flight Plan Generation	Utilization Dependent on Airspace Complexity // Implementation 1997/98
2 d	Dynamic User Preferred Routes	Phased implementation user supplied single reroute LAX-SYD-LAX LAX-AKL-LAX NAN-YVR-NAN NAN-LAX-NAN	CNS /ATM Approved Aircraft, with FMS (IRS) CPDLC AOC Data Link Direct Flight Plan Uploads	CPDLC AIR/GND DATA LINK AOC/ACC Data Communication Flight Plan Generation ATC Traffic Situation Display.	Utilization Dependent on Airspace Complexity // Implementation 1998-99
2 e	Free Flight	Not currently being evaluated	Autonomous Aircraft	ATM system that is interoperable with Autonomous Aircraft	Activities in other areas being monitored for future consideration.
3 VERTICAL SEPARATION					
3 a	1000' to FL290, 2000' above FL290	All routes	ICAO SARPS	ICAO SARPS	// Implemented
3 b	2000' Step Climb above FL290	All routes			Pending general use of RVSM above FL290 // Implement post 2000
3 c	Cruise Climb	All routes	FMS (IRS) CPDLC ADS ACAS	CPDLC ADS	// Implemented today through block altitudes assignments.
3 d	1000' Crossing Traffic above FL290	Selected (TBD) crossing points	FMS RTA ACAS with Traffic Situation display RVSM Certification	ATC Traffic Situation Display	Pending general use of RVSM above FL290 // Implement post 2000 ADS may be required for ATM

169

Table 11.4 (continued)

3 e	1000' Vertical Separation between FL290 and FL450	Not currently being evaluated.	RVSM Certification	ICAO SARPS ATC Traffic Situation Display	Activities in other areas being monitored for future consideration. // Implementation post 2000.
4 LONGITUDINAL SEPARATION					
4 a	80nm	All routes	RNAV Capability		// Implemented
4 b	50nm	All RNP 10 routes	RNAV or FMS DCPC Voice or Data RNP 10 Approval	DCPC Voice or Data	30 Min Reporting Requirement Available today // Implementation 1997
4 c	30nm	Currently being evaluated.	Indicative requirements: RNAV or FMS DCPC Voice or Data RNP 4 Approval ADS	DCPC Voice or Data ADS	15 Min Reporting Requirement ADS required by States for ATM. // Implementation TBD
4 d	Less than 30nm	Not currently being evaluated	Indicative requirements: RNAV or FMS DCPC Voice and Data RNP Approval ADS	DCPC Voice and Data ADS	// Implementation TBD
4 e	10 min	All routes	RNAV Capability		// Implemented
4 f	7 min	Not currently being evaluated	RNAV or FMS DCPC Voice or Data RNP 10 Approval Accurate Time	DCPC Voice or Data Accurate Time	// Implementation TBD
4 g	4 min	Not currently being evaluated	Indicative requirements: RNAV or FMS DCPC Voice or Data RNP 4 Approval Accurate Time	DCPC Voice or Data Accurate Time	// Implementation TBD

Table 11.4 (continued)

5 LATERAL SEPARATION				
5 a	60nm		RNAV Approval	// Implemented
5 b	60nm	All RNP 10 routes	RNAV or FMS RNP 10 Approval	DOC 7030 Amendment. Available today // Implementation 1997
5 c	30nm	Currently being evaluated	RNAV or FMS RNP 4 Approval DCPC voice or data ADS	DCPC Voice or Data ADS	15 minute position reporting required // Implementation TBD
5 d	Less than 30nm	Not currently being evaluated	RNAV or FMS DCPC Voice and Data RNP Approval ADS	DCPC Voice and Data ADS	// Implementation TBD

Table 11.5

AREA: ASIA - NORTH AMERICA (via North Pacific and the Russian Far East)

Current Description:	Lead Aircraft: B/747/767/777, MD11, A340, IL96
Generally PANS RAC/DOC 4444 SUPPS/DOC 7030 Separation: 100nm/10min/2000' Fixed ATS route/2000' step climb/1000' 500m In RFE. Composite in NOPAC	Note: For each particular operational enhancement, there will be a need for the airlines and the ATS Providers to review existing procedures to identify what new requirements are needed prior to operational implementation.
Communications: Voice: VHF/ SATVOICE/ HF/HF Intermediary CPDLC: VHF/SAT **Navigation:** FMS/Inertial/GNSS **Surveillance:** Procedural/Radar/ADS	

CODE	ATM OPERATIONAL ENHANCEMENTS	SYSTEM IMPLEMENTATION	REQUIRED FUNCTIONS - AIR	REQUIRED SERVICES - GROUND	NOTES // TARGET IMPLEMENTATION DATE
1 RNAV ROUTES					
1 a	RNAV Routes	Most routes	RNAV Capability		// Implemented
1 b	Flexible Tracks	Canadian and Alaskan transitions to RFE/NOPAC	FMS or RNAV	Track Generation/ Distribution	NOPAC – North Pacific Tracks // RFE – Russian Far East Implemented
1 c	RNAV Parallel one way Route Structures	Selected NOPAC routes	RNAV capability		// Implemented
2 CNS/ATM ROUTES					
2 a	Fixed Routes	Selected routes in NOPAC and RFE.	CNS/ATM Approved Aircraft	May require DCPC (Voice or Data)	Extensions into China to be evaluated by China. RFE is evaluating five CNS routes (RNP 4) // Implementation 1997
2 b	Flexible Routes	Not currently being evaluated	CNS /ATM Approved Aircraft	May require DCPC (Voice or Data)	// Implementation TBD

Table 11.5 (continued)

2 c	Dynamic Re-Route	Transition from Canadian and Alaskan tracks currently being evaluated	CNS /ATM Approved Aircraft, with FMS (IRS) CPDLC AOC Data Link Direct Flight Plan Uploads	CPDLC AIR/GND DATA LINK AOC/ACC Data Communication Flight Plan Generation	// Implementation TBD
2 d	Dynamic User Preferred Routes	Transition from Canadian and Alaskan tracks currently being evaluated	CNS /ATM Approved Aircraft, with FMS (IRS) CPDLC AOC Data Link Direct Flight Plan Uploads	CPDLC AIR/GND DATA LINK AOC/ACC Data Communication Flight Plan Generation ATC Traffic Situation Display.	// Implementation TBD
2 e	Free Flight	Transition from Canada / Alaska currently being evaluated	Autonomous Aircraft	ATM system that is interoperable with Autonomous Aircraft	US Flight 2000 Program to demonstrate Free Flight // Implementation TBD
3 VERTICAL SEPARATION					
3 a	1000' to FL290, 2000' above FL290	All routes except as noted	ICAO SARPS	ICAO SARPS	// Implementation China, DPR Korea and Russia TBD
3 b	2000' Step Climb above FL290	NOPAC			// Implemented
3 c	Cruise Climb	In Anchorage FIR	FMS (IRS) CPDLC ADS ACAS	CPDLC ADS	//Implemented in US airspace traffic permitting. Does not require CNS/ATM capability.
3 d	1000' Crossing Traffic above FL290	Not currently being evaluated	FMS RTA ACAS with Traffic Situation display RVSM Certification	ATC Traffic Situation Display	
3 e	1000' Vertical Separation between FL290 and FL450	NOPAC	RVSM Certification	ICAO SARPS ATC Traffic Situation Display	// Implementation date TBD. Most likely post 2000.

Table 11.5 (continued)

4 LONGITUDINAL SEPARATION				
4 a	80nm	Not currently being evaluated	RNAV Capability	
4 b	50nm	All routes in NOPAC	RNAV or FMS DCPC Voice or Data RNP 10 Approval	DCPC Voice or Data	May require ADS // Implementation 1998
4 c	30nm	All routes in NOPAC currently being evaluated	Indicative requirements: RNAV or FMS DCPC Voice or Data RNP 4 Approval	DCPC Voice or Data	May require ADS // Implementation TBD
4 d	Less than 30nm	Not currently being evaluated	Indicative requirements: RNAV or FMS DCPC Voice and Data RNP Approval ADS	DCPC Voice and Data ADS	Implementation TBD
4 e	10 min	All routes	RNAV Capability		Implemented with Mach Number Technique
4 f	7 min	Not currently being evaluated	RNAV or FMS DCPC Voice or Data RNP 10 Approval Accurate Time	DCPC Voice or Data Accurate Time	// Implementation TBD
4 g	4 min	Not currently being evaluated	Indicative requirements: RNAV or FMS DCPC Voice or Data RNP 4 Approval Accurate Time	DCPC Voice or Data Accurate Time	// Implementation TBD

Improving capacity

Table 11.5 (continued)

5 LATERAL SEPARATION					
5 a	60nm	Not currently being evaluated	RNAV Approval		
5 b	60nm	All routes in NOPAC	RNAV or FMS RNP 10 Approval		// Implementation 1998
5 c	30nm	All routes in NOPAC	RNAV or FMS □ RNP 4 Approval DCPC voice or data	DCPC Voice or Data	// Implementation TBD
5 d	Less than 30nm	Not currently being evaluated	RNAV or FMS DCPC Voice and Data □ RNP Approval ADS	DCPC Voice and Data ADS	// Implementation TBD

175

References

ICAO, Special Committee on Future Air Navigation Systems : Fourth Meeting Montreal, 2-20 May 1988, *Reprint (DOC 9524, FANS/4)*.

ICAO, Report on the 10th Air Navigation Conference : Montreal, 5-20 September 1991, *(DOC 9583, AN-CONF/10)*.

ICAO, Special Committee for the Monitoring and Coordination of development and Transition Planning for the Future. Air Navigation System (FANS Phase II), Report of Fourth Meeting, Montreal 15 September- 1 October 1993, *DOC 9623, FANS (II)/4*.

ICAO, Asia and Pacific Office, *Asia/Pacific Regional Plan for the new CNS/ATM Systems*.

The author

Professor Brian O'Keeffe is Special Technical Adviser to the Chief Executive officer of Airservices Australia and Chairman CNS/ATM Implementation Coordination Sub group of the ICAO Asia/Pacific Regional Planning Group. He has had a long and distinguished career in the air traffic field, with his significant contribution to the development of FANS activities being clearly shown in this chapter.

12 The new IATA International Passenger Liability Regime

Lorne S. Clark

On 28 May 1996 British Airways opened a new chapter in the annals of international air law by becoming the first signatory to the 'Agreement on Measures to Implement the IATA Intercarrier Agreement on Passenger Liability'.

This event marked the initiation of what can best be termed an historic 'rite of passage' from the old Warsaw Convention limited liability system initiated in 1929, amended at The Hague in 1955 and supplemented by means of the Montreal (Intercarrier) Agreement of 1966. The new IATA unspecified limits regime stands in bold contrast to the system it supplants.

In particular, the liability provisions governing carriage by air have been irrevocably altered. True, it will take more time to gain the degree of universal acceptance required to ensure global implementation of the modernised regime. But this is not only inevitable, given the commitment on the part of major carriers around the world, but it is assured by the unyielding determination to provide 'unspecified limits' for international passengers on the part of the Japanese and United States Governments and the European Union.

A few somewhat reluctant partners and ivory tower academicians might be forgiven for not realising that the new liability regime flight has already taken off, while they are still debating the merits of increasing outdated and discredited fixed limits. After all, the international aviation community has been arguing about increasing the Warsaw/Hague limits since the ICAO Guatemala City Air Law Conference of 1971, without success. Who could ever have imagined that in less than two years since the IATA Airline Liability Conference (ALC) in Washington DC in June 1995, a groundbreaking, modern intercarrier liability regime would be adopted by airlines carrying more than 80 % of scheduled international air transport?

The momentum is gaining, the old order is gone, and governments are only now scrambling to try to catch up with the airlines by attempting, through ICAO, to develop an entirely new inter-governmental treaty framework for the unspecified limits regime.

Let us examine some of the history behind the remarkable developments of 1995-97. On 22 February 1995, the United States Department of Transportation (DOT) issued Immunity Order Number 95-2-44 authorising IATA to conduct intercarrier discussions 'to secure the important public benefit of a liability regime that reflects contemporary standards of compensation'. Having much earlier received the requisite 'negative clearance' from the European Commission, on 1 September 1993, IATA immediately began a flurry of focused activity. Firstly, the Secretariat established a Special Legal Working Group to plan and organise the convening of the Airline Liability Conference. Under the able chairmanship of Air Canada's then Senior Vice President and General Counsel, Cameron DesBois, it met twice in Washington in April and in Atlanta in May 1995. As a result the process and procedures for the ALC were developed and refined, and relevant documentation prepared by the IATA Secretariat. Taking into account the conditions and 'guidelines' set out in the DOT Order, the agreed approach was basically straightforward:

- invite the entire global international airline industry to participate in the ALC;

- generally preserve the uniform Warsaw Convention framework;

- acknowledge that the liability status quo in the Warsaw system was untenable;

- modernise passenger liability limits through the intercarrier agreement mechanism;

- develop the overall package in a manner likely to receive requisite government approval.

The ALC met in Washington 19-23 June 1995 and was attended by 67 carriers from around the world, 6 regional airline associations, 3 other industry associations, and representatives of ICAO, the European Commission and the US Departments of Transportation, Justice and State. Having received the honour of being elected Chairman of the Conference, the author stated in his opening remarks on 19 June:

There is a significant challenge before us here. Let me say I firmly believe all of us have a responsibility to represent not only the entity that sent us to Washington, but the interests of the industry at large. Much as members of a 'constituent assembly' or a constitution-writing group, we have to look beyond narrow parochial interests and seize the moment to serve our carriers, the industry at large and the travelling public. The challenge is to find and agree on a balanced solution to the liability issues.

After the usual round of opening statements and agreement on an appropriate work programme, the ALC quickly approved the establishment of two Conference Working Groups and organised its agenda in a manner to allow for regional and other interest group meetings on the margins of the main discussions. At the end of the session on 23 June 1995, the Conference approved a Final Report setting out a number of specific conclusions. These included mandates for the IATA Secretariat to prepare the text of an intercarrier agreement and a 'means to secure complete compensation' for international passengers, and for the chairman to appoint special subcommittees to assist in carrying out these tasks. The eventual intercarrier agreement was to be submitted to the IATA Annual General Meeting (AGM) for approval at the end of October 1995 and then to governments as required.

The two subcommittees were duly constituted; however their members decided to merge into a single joint body under the chairmanship of the author. This group met twice, in London in July and in Washington in August 1995. As a result of its deliberations, and review and finalisation by the original Special Legal Working Group, an 'umbrella' Intercarrier Agreement on Passenger Liability (IIA) was drafted and agreed, leaving the particular implementation provisions to be negotiated at the next stage.

The author then undertook a series of consultations around the world, including with the then Orient Airlines Association (OAA, now Association of Asia Pacific Airlines - AAPA) in Manila, the Arab Air Carriers Organization (AACO) in Cairo, the Association of African Airlines (AFRAA) in Nairobi, the Association of European Airlines (AEA) in Brussels, the Air Transport Association of America (ATA) in Washington, and the European Commission in Brussels, as well with members of the International Association of Latin American Air Transport (AITAL) and a number of individual carriers, in advance of the October 1995 IATA AGM.

Building on the support thus secured, the IATA Intercarrier Agreement was unanimously endorsed by a Resolution of the IATA AGM in Kuala Lumpur, and opened for signature at an initial symbolic signing ceremony on 31 October 1995. The objective was to bring the IIA, together with appropriate implementing provisions, into force by 1 November 1996. Chief Executive Officers of 12 carriers, representing all of IATA's geographic regions, were the initial signatories. With the impending expiry of DOT Immunity Order 95-2-44 on 31 December 1995, IATA requested and received an extension of immunity to 1 April 1996, later extended to 1 July 1996 to facilitate development of the IIA implementation provisions.

The IIA umbrella agreement commits its signatories:

> To take action to waive the limitation of liability on recoverable compensatory damages in Article 22 paragraph 1 of the Warsaw Convention as to claims for death, wounding or other bodily injury of a passenger within the meaning of Article 17 of the Convention, so that recoverable compensatory damages may be determined and awarded by reference to the law of the domicile of the passenger.

The limitations to be waived have of course been in effect since 1929 and were originally designed to protect the interests of an infant airline industry. As well-known US plaintiffs' attorney Lee Kreindler so succinctly stated at the International Aviation Law and Insurance Seminar in London in November 1995, following the author's presentation on the IIA, 'the devil is in the details'. The particular 'devil' concerned what kind of action an airline must take 'to waive the limitation of liability'. Basically, it was generally agreed this could be accomplished by the following:

- tariff filings already accepted by government (as done by the Japanese carriers in 1992);

- conditions of carriage/new tariff filings submitted to government (as done by several carriers in 1996-97);

- an implementing intercarrier agreement giving effect to the IIA (binding on signatories, to be filed separately with government);

- government-imposed implementation (potential example: December 1995 European Commission).

Working principally with the members of the original Legal Working Group, now expanded to include representatives of the growing list of IIA signatories, the IATA Secretariat initiated an intensive effort at the beginning of 1996 to elaborate acceptable provisions to implement the IIA. At the same time airlines and their regional associations around the world were engaged in urgent discussion and debate about the best means to do this, and provided continuing guidance to the Secretariat.

On 1 February 1996 in Miami, the expanded Legal Working Group approved, with the US carrier representatives abstaining - on the grounds the new accord would not be acceptable to the US DOT - a draft text of IIA implementing provisions, subject to review with interested governmental authorities. A high-level meeting with DOT took place in Washington on 14 February 1996, also attended by representatives of a number of non-US carriers and regional airline associations, at which the proposed implementation text was explained and discussed. In addition, there were parallel discussions with the European Commission and other government representatives.

On 12 March 1996 the DOT General Counsel sent the Secretariat a letter commending IATA's efforts but calling for improvements to the Miami text in order to help ensure approval by US authorities. In particular, DOT requested reconsideration of a) the 'optional' aspect of the provision regarding determination of damages by reference to law of the passenger's domicile, preferring that it be mandatory; and b) the possibility of incorporating a specific right for the claimant to litigate in the territory of the passenger's domicile or permanent residence, when this forum is not available under the Warsaw Convention (the addition of a so-called 'fifth jurisdiction' to the four Warsaw Convention Article 28 fora).

A final meeting of the Legal Working Group was held in Montreal on 3 April 1996 to review the Miami draft, taking into account the discussions with and suggestions put forward by DOT. After extensive debate, the DOT proposals were rejected and the Miami text, with minor editorial amendments, was approved - this time with the support of the US carrier representatives. The document was entitled 'The Agreement on Measures to Implement the IATA Intercarrier Agreement' (MIA). As adopted, it enshrines the principle of full 'recoverable compensatory damages', provides for SDRs 100,000 'strict' liability except where carriers and governments may specifically agree otherwise, reiterates the *option* on the part of the carrier to accept domiciliary law for calculation of damages, does *not* add to the Article 28 fora available to a plaintiff for litigation but does offer carrier signatories other options not inconsistent with the MIA provided they are in accordance with applicable law.

Transcending both the IIA and the MIA, the *law of the domicile* issue has perhaps given rise to the most comment, often simply because of confusion with the 'fifth jurisdiction'. Making the 'quantum' leap from liability limits often as low as US$ 10,000 (under the unamended Warsaw Convention) to unspecified limits, has undoubtedly posed a serious psychological and financial challenge to many airlines, especially small and medium-sized carriers. In order to possibly mitigate their exposure while enshrining the new unspecified limits approach, some carriers prefer an option to have the courts calculate the damages according to the law of the passenger's domicile or permanent residence, that is the law most closely connected with the victim. The MIA spells this out much more clearly and definitively than the IIA 'umbrella agreement' i.e. that the law of the domicile is an *option available to the carrier*.

With respect to the 'fifth jurisdiction' i.e. permitting a claimant to litigate in the territory of the passenger's domicile or permanent residence, the

IATA Secretariat and many airlines take the position it would be contrary to and inconsistent with the Warsaw Convention to attempt to create this by intercarrier agreement when it is not a forum available under the Convention. For example the European Commission (if it is, in fact, within EU competence and if the Member State governments so permit) may impose it on *European Union* carriers and the US authorities may impose it on *their* airlines, but, IATA maintains, neither authority can impose the fifth jurisdiction on third country airlines without amending the Convention. In any case, as noted above, the MIA does not purport to add to the treaty jurisdictions.

The European Commission accepted the IIA and MIA on 26 November 1996. Following a substantive debate between IATA and the US DOT, reflected in the latter's 'Show Cause' Order (SCO) of 3 October 1996, IATA's massive legal objections to the SCO and to the proposed conditions to approval of the IIA/MIA regime and, finally, the DOT's formal approval of the IIA and MIA on 8 January 1997, both intercarrier agreements came into force. For the record and in accordance with its terms, the Director General of IATA declared the MIA 'effective' on 14 February 1997.

At the time of writing, mid-March 1997, the IIA had attracted 83 signatories and the MIA had 51 signatures, with both numbers expected to grow in the months ahead.

It is important to underscore, once again, that the Warsaw/Hague instruments have not been supplemented for three decades, that is since IATA negotiated the 1966 Montreal Agreement (which covers only service to, from, and with an agreed stopping place in, the United States) under the 'gun' of US denunciation of the Warsaw Convention at that time. The 1966 Agreement, of course, did not amend those instruments, but took advantage of the possibility of the special agreement mechanism to increase passenger liability limits and to voluntarily provide for the waiver of Article 20(1) defences on the part of carriers. (The then US Civil Aeronautics Board (CAB) - predecessor of the DOT - later made signature of the Montreal Agreement mandatory for carriers serving the US, failing which they would be 'deemed' to be a party in any case. This remains the current situation, although IATA is now working to orchestrate withdrawals from the 1966 Agreement, which has effectively been superseded by the IIA/MIA regime).

IATA has long maintained that the Warsaw system is seriously out of date and that the divergent liability limits established around the globe were rooted in the past and did not reflect contemporary community standards in much of today's world. Because of this, the applicable limits of compensation have often been successfully attacked and set aside, by US courts in particular, under the guise of the carrier being guilty of 'wilful misconduct', with attendant exposure to unlimited liability under the Warsaw treaty.

Efforts by ICAO and Government representatives in Guatemala City and Montreal, in 1971 and 1975 respectively, to amend the Warsaw system failed. These had been directed to providing for an unbreakable 'cap' on airline liability and, in the 1975 Montreal Additional Protocol3 (MAP3), to allow for a passenger-funded but government-administered supplemental compensation scheme above and beyond the capped amount, where necessary or desirable (primarily in the US). Barely a year after the Montreal Conference, the proposed 'cap' of SDRs 100,000 was already considered inadequate by the US Government - and the amendments to the treaty never came into force for lack of sufficient ratifications, especially the absence of 'advice and consent' on the part of the US Senate.

More recently, the 1993-94 attempts by the US Administration to develop a new liability regime under the Warsaw Convention, including the promoting an updated MAP 3 also failed.

IATA Members, reiterating their firm commitment to enhancing the liability benefits to the travelling public, nevertheless must operate within the constraints affecting airlines, which are legally not able to amend international treaties; this responsibility rests with ICAO and governments. As they embark, once again, on this task, they should note that since February 1995, IATA and its Member carriers have struggled with the fact that the means of achieving this are far from simple. Indeed the Secretariat has been guided by the following, occasionally conflicting, needs:

- to harmonise the positions of major carriers with that of small and medium-sized airlines, all of them having to put in place appropriate insurance coverage in a relatively short time frame;

- to rationalise the needs of carriers already possessing insurance covering 'unspecified limits' and those offering only Warsaw, Hague, or other fixed limits, some mandated by governments (with the amounts largely unbeknownst to their passengers);

- to preserve the universality of the Warsaw Convention framework, while accommodating inevitable differences in conditions of carriage around the world.

In short, IATA has had to try to overcome all the impediments which have prevented Warsaw reform for the thirty years since the 1966 Montreal Agreement, despite concentrated effort, personal commitment by diligent individuals and, what seemed at various times, a promising environment. The degree of success is indicated in Table 12.1 which shows over one hundred Carriers were signatory to the Agreement as at 24 January 1998. The present challenge to ICAO should thus not be underestimated.

There is a slow but growing recognition of how truly historic the modernisation of the Warsaw system by IATA really is. It cannot be overemphasised that the old Warsaw/Hague liability world is dead and is in the process of being interred. There will be no going back to artificial specified liability limits. The main task for IATA and the international air transport industry in the years ahead will be to try to harmonise as much as possible, within a common framework, the *means* of IIA implementation and, perhaps, to enhance even further the benefits to the travelling public. This will be the central challenge - to ensure the new global approach is just that, *global* in scope, despite possible regional variances to conform to local laws and regulations or the perceived needs of particular environments. What ICAO and governments, with the best will in the world, have been unable to do in the more than four decades since the adoption of The Hague Protocol in 1955, the airlines have accomplished, and in less than two years. In the end, this will be to the benefit of the carriers, the passengers, and to the insurers as well.

Governments have now been provided by IATA with a 'flight plan' to follow suit. It is up to them to move expeditiously to enshrine the new regime in binding international law, the task on which ICAO has just embarked. In this venture it can count on the strong support of IATA and its more than 250 member airlines. However, a word of advice is proffered: reform Warsaw/Hague as necessary, but *only* as necessary, or risk more 'paper' amendments which will lie on the shelf next to the Guatemala City Protocol and Montreal Additional Protocol 3, providing interesting reading for academics and lawyers but no benefits whatsoever to the travelling public, or the air transport industry!

The author

Lorne S. Clark is the General Counsel and Corporate Secretary of the International Air Transport Association (IATA).

Table 12.1
List of carriers signatory to the IATA intercarrier agreement on passenger liabilty as at 24 January 1998

Aer Lingus plc	Continental Airlines Inc.
Aerolineas Argentinas S. A.	Continental Express
Aeromexpress	Continental Micronesia
Aerovías de México, S.A. de C.V. (Aeromexico)	Croatia Airlines
	Crossair
Air Afrique	Delta Air Lines, Inc.
Air Aruba	Deutsche BA Luftfahrtgesellschaft mbH
Air Baltic Corporation SIA	
Air Canada	Deutsche Lufthansa AG
Air Exel Commuter	Egyptair
Air France	Emirates
Air Mauritius	Eurowings Luftverkehrs AG
Air New Zealand	Finnair OY
Air Pacific Limited	Garuda Indonesia
Air UK Group Limited	GB Airways
Air Vanuatu	Hawaiian Airlines
Alaska Airlines	Heli Air AG
Alitalia	Heli-Linth AG
All Nippon Airways Co., Ltd	Iberia
Allegheny Airlines, Inc.	Icelandair
America West Airlines, Inc.	Interimpex-Avioimpex
American Airlines	Japan Air Charter (JAZ)
American Trans Air, Inc.	Japan Air System Co. Ltd
Asiana	Japan Airlines Co. Ltd.
Augsburg Airways GmbH	Japan Asia Airways (JAA)
Austrian Airlines	Jet Airways (India) Pvt Ltd.
Avianca	Kenya Airways
Azerbaijan Hava Yollary	Kiwi International Air Lines
Braathens S.A.F.E.	KLM Cityhopper B.V.
British Airways p.l.c.	KLM Royal Dutch Airlines
Canadian Airlines International	Korean Air Lines Co., Ltd.
Cathay Pacific Airways Ltd.	LAPSA Líneas Aéreas Paraguayas
Central Mountain Air Ltd	
Cimber Air A/S	Lauda Air Luftfahrt AG
Compagnie Air France Europe	Luxair

Developing the Future Aviation System

Maersk Air A/S Maersk Air Ltd. Malaysia Airlines Malev – Hungarian Airlines Public Ltd. Co. Martinair Holland N.V. Midwest Express Airlines, Inc. Northwest Airlines, Inc. Pakistan International Airlines (PIA) Piedmont Airlines, Inc. Polskie Linie Lotnicze - Polish Airlines PSA Airlines, Inc. Qantas Airways Limited Reeve Aleutian Airways, Inc. Regional Airlines Royal Air Maroc SABENA Saudi Arabian Airlines Corp. Scandinavian Airlines System (SAS)	Singapore Airlines Ltd. South African Airways Swissair TACA TAP Air Portugal TAT European Airlines Trans World Airlines Inc. (TWA) Transavia airlines C.V. Transbrasil S/A Linhas Aéreas Trinidad & Tobago BWIA International Türk Hava Yollari A.O. (Turkish Airlines) Tyrolean Airways - Tiroler Luftfahrt-AG United Airlines UPS Airlines USAir, Inc. Varig S.A. VIASA

13 Developments in aircraft interior design

Carole Favart-Andrieux

Introduction

Even though technological evolution in the aviation field has taken giant steps forward, there has been less effort concerning the comfort of passengers and life aboard the aircraft; especially when considering 'economy class'. However, the last ten years have seen many improvements.

As the aviation sector generally excites more passions than other industries, performance is often the main feature which is aimed for (e.g. Marcel Dassault's comment that 'a beautiful aircraft flies better'). Consequently, the passenger may sometimes be considered merely a body which participates in the commercial air transport enterprise.

With the end of monopolies for some airlines, and the general liberalization of the aviation sector, increased competition has been a logical consequence with the result that aircraft operators must more than ever learn to seduce the passenger. Most of the main manufacturers have addressed this requirement in more or less similar ways; with the result that two key questions arise:

• How can an aircraft manufacturer persuade an airline to select their aircraft rather than other ones?

• How can an airline use the aircraft to keep their present clients and attract new ones as well?

Some general points on the existing problems can be made without being specific to particular constructors. Too often the colour schemes are dull and the cabin atmosphere can be rather aseptic. The only parts providing a more personal atmosphere are the seats, the carpet and sometimes a discreet motive on the porthole panels. In fairness, the cabin structure creates a difficult situation and this is exacerbated in the economy class. Here, the comfort is often very spartan and only a slight inclination of the seat is possible - if the passenger does not have very long legs!

Other discomforts depend on the difference of perception by the passengers with respect to such things as cigarettes, temperature, ambient light, smells and noise. Those, which could be tolerable on middle distance flights, can become intolerable on a long haul.

Fortunately, there are many modern approaches which can be made to improve the comfort of flight. These will be covered in this chapter, under broad terms such as the impact of colour schemes, the effects of industrial design, and the scope for the architecture of the cabin interior and the application of ergonomic principles.

Impact of colour schemes

This aspect can encompass a wide range of techniques (e.g. colorimetry, psychometry) which can be employed, often in concert, in order to avoid the choice of a colour scheme being too subjective. Indeed, colour plays a major sociological role in such matters as ergonomics, functionality, safety, comfort, the movement of passengers and the provision of information.

By balancing the artistic and technical work, specialists such as Chevreul, Goethe, Itten and more recently Filacier, have achieved some very concrete results. Hence the methodology can improve the interiors of aircraft in a meaningful way and at the same time solve some vestigial problems.

However, in using colour techniques (that is, the application of colorimetry psychometry) there is still much to learn and further developments will arise. There is scope for numerous individual options which can help to personalise the travel experience. In fact, researches show that the use of colour can be subjectively abstract, like a language or code, and have physiological implications. In this respect the human responses can be akin to hearing music.

The aesthetic aspects of colours can be considered with respect to a triple point of view:

- sensibilities and optic (i.e. impression of the colour)

- psychic (i.e. expression)

- intellectual and symbolic (construction of the colour)

These have been expressed by Itten, Professor at the Bauhaus, as well as Goethe, Bezold, Chevreul, Hölzel and Fillacier. The aspects can be represented as psychological and functional impacts, the messages imparted, the meaningfulness and the ambience involved.

The psychological impact

This important aspect operates at two levels.

The psychological level

Goethe thought that it was necessary to use the colours as a function of their influence on the mind. For instance hot colours such as red and yellow can invoke some quick, rapid or intense feelings, whilst cold hues of blue, green and blue-red can produce agitation and anxiety.

Experiments have demonstrated that there was no interrelationship between the psychic impact and the physiological impact; nevertheless the sensations are acknowledged to have an influence on the psychology of an individual. This could be a positive aspect or negative as, for instance, a soothing blue, could have too much affect in inspiring boredom or even a feeling of cold weather numbness rather than being refreshing!

It is also important to consider the cultural context of different ethnic groups as the impact of colours varies in respect to age, geographical place etc. Thus their use in aircraft needs to take account of passenger profiles.

The physiological level

The first colourful perceptions of mankind seem to have been associated with the human body; thus the red of blood and the colours of other bodily fluids. The interest for a cabin interior designer is whether these background influences could provoke a particular view of colour in an individual and so provide a broad sense of comfort and a sensation of security. The same affects are also associated with the nature of the surface materials (e.g. soft, smooth, transparent, opaque, thick or thin) together with the shapes involved.

The functional impact

The psychological aspect must also be understood in order to create not only a suitable ambience but also to stimulate emotions, reactions and other responses of a passenger. Thus in restricted spaces it will be possible to generate positive reactions of well being through harmonious design of colours, lights and music.

Developing the Future Aviation System

It must also be remembered that the design needs to take into account the responses of different groups. For instance, females may prefer a different atmosphere to a traditional 'masculine' one.

The messages imparted

If the aircraft is in international use a rational use of the design can help in improving messages which are needed for such matters as access, indications of dangers, providing news, directions or indicating stages of a flight. This process is of course already used in many areas such as maintenance as well as on-board services, which might utilise the different brightness of lights, or colour changes. Naturally, these aspects apply to airports as well as in aircraft and it is certainly desirable if a certain amount of harmonisation could be achieved as the passenger sees the airport and the flight as simply different stages in their journey.

The meaning of colours

As with music, colours have implications of character, values (positive and negative) and significance, which vary, across the continents. For instance, white can represent mourning in Asia, as opposed to black in most Western countries. Thus, some airlines might find it beneficial to have some accoutrements, which are moveable in order to provide appropriate alternatives available for specific flights.

The meaning of the colours will also be influenced by the special use of shapes which can be useful in providing a 'visual illusion' to solve the problem of enclosed and restricted spaces. The different techniques would include the height, width and depth of fields together with the opacity and transparency. Use of these 'lines of strength' can be effective in promoting dark, hot or cold qualities.

Itten has suggested that these effects can be categorized using seven contrasts together with an eighth category concerning materials.

- Contrast of the colour itself:

- Clear - dark contrast

- Hot - cold contrast

- Complementary contrast

- Simultaneous contrast

- Contrast of quality

- Contrast of quantity

- Contrast with the material.

The meaning of colours is also interconnected with their symbolism. This is a vast topic but it is useful to mention that the colours of today vary greatly from those used in previous times and often have quite different symbolic meanings.

The ambience created

It is important to avoid producing a monotonous ambience around a passenger. Thus, without him or her changing places a careful use of the lights can make it possible to 'sculpt' the space differently according to the phase of the flight e.g. relaxing for meal-times or lively when preparing for landing. The scope for such variations has increased tremendously due to the development of new technologies, such as fibre optics, in this area.

Industrial design

For some years key improvements focused on personal comforts of the passenger, e.g. Flat screen video, telephones etc. Now it is necessary to address matters such as lighting systems, the principles of ventilation and how improvements can be achieved through the use of new technologies and new materials. Thus it is important for the industrial and commercial designers to co-ordinate all aspects from research to marketing while producing the final product. This means giving consideration to all the key components.

Modern technology offers a great scope of possibilities in constructing the interiors of aircraft although economic constraints mean that a balance has to be achieved between capital investment, market needs and the industrial design. Fortunately, the human needs are now much better understood due to the integration of psychology, sociology, ergonomics and the broader realms of Human Factors studies. All these components, together with aesthetic considerations, support an improved ability to meet the more personal expectations of passengers.

Developing the Future Aviation System

The stages of design

It is useful to consider the design process as covering three essential stages. The first stage is the research phase during early studies when principles and concepts are elaborated. The results of the studies enable the designer to make the technical, ergonomic, economical and aesthetic, fundamental choices. In the second phase all the options of structure and of shape are explored resulting in plans and models being made. Finally, the third phase, aimed at industrialisation, deals with the details, and aims at verifying the feasibility and the viability of the techniques employed. Naturally, it is the outcomes of the third stage which produce results for the passenger. More significantly for the airline, new developments are soon apparent.

Short term impact

Design can have a major impact on the cockpit as this is a place of work and consequently a great deal of research and study is put into this area for safety and other reasons. In return, it is useful to commence cabin design with a consideration of why the passengers are there - using perhaps the initial responses of passengers as they board the aircraft.

One of the first features which a passenger meets is the seat as this is in direct contact with the body of the passenger. Unfortunately, it is a fact that it is only the seats of first and business class which seem, to the economy passenger at least, to have benefited from recent studies.

In first and business classes extra comfort, and possible transformation to bed-rests, are provided. In many cases, extra services such as television and communications are individually available. It is possible that even more extra comfort might be added in future by seats which could also move on a vertical axis. However, due to penalties of extra weight it is still questionable whether such improvements would be possible.

Such penalties though do seem to be feasible in terms of further developments, which will take place towards providing better sleeper-beds for long haul flights. Naturally, such improvements will necessarily call for appropriate ways of securing the sleeper in the berth together with improved personal heating and ventilation. As always, a balance has to be made between the costs of providing personal comfort and the costs of enhanced services.

It is common practice for many of the first class and business class features to gradually devolve down to economy class. But, there are a number of ways in which access to the improved level of service can be achieved by designing the layout of the cabin. For instance, there could be one telephone for three persons and sleeper beds, in special areas, could be booked for short sections of the journey.

The passenger in control

With improved technology and complicated switching systems for levels of light, ventilation, sound etc. it will be important to simplify the procedures and allow passengers to be in control of the mix that they require rather than being overcome by a multitude of possibilities. However, the results of one persons personal freedom might result in unpleasantness to another passenger, especially due to the constricted space. There will be a need for some commands to be blocked (e.g. perhaps with overactive children) or coded where some functions are not needed on certain flights.

Passenger communications

It has to be recognized that a passenger in an aircraft is now part of the Global Village. That is, two way communication, with increasingly all forms of information transfer, is now available almost everywhere and anywhere. Aircraft are no exception with the result that passenger communications is a fast growing sector of the available services.

Besides telephones, fax, video films and live television the modern services include access to the Internet. Certainly, the extra scope for business or pleasure activity, in the individual passenger seat, strengthens the case for providing all the communication services, which are available in the office or home.

However, it must be recognized, that the passengers are in a restricted area and other people are closer than would normally apply in the office or home. Thus, there could be scope for a new approach which offers special functional areas that provide special services, such as private cubicles, a variety of computer capabilities (just bring the disks and not the lap-top), different sizes of computer screen according to the use, and flexible charging systems! The same arguments would apply for electronic games, films, video phones and video conferencing.

Mobility in the aircraft

On long-haul flights passenger naturally want to move around in order to talk with friends, colleagues or simply stretch their limbs. Unfortunately for them the aircraft is flying at a great speed through an often-difficult environment and so it is necessary to provide protection and comfort devices for the passenger.

Modern materials offer a wide range of possibilities to add even more extra comfort. These include components which can slide or roll, suction pads which hold together hard surfaces, and fabrics which can cling together by the use of Velcro. In fact the improvement of thin materials with soft surfaces and lack of roughness are a great boon to designers and present studies on 'memory shapes' will take the process further.

The architecture of the interior

As mentioned earlier, an aircraft is a single entity and consequently all the domains of an aircraft need to be coordinated. This includes areas such as the interior space, the full aspects of the circulation, the different galleys and the features concerning harmonization such as ambience, colours and materials.

At present the differences offered in the different cabin classes might be perceived simply in terms of the cabin catering, the size and quality of seating, a certain subtlety in colour schemes and more space in the passageways.

Unfortunately, for the designer, there is a main constraint, in considering new cabin designs, caused by the airframe structure whose fuselage shape is one of a 'tube' with a repetitive pattern of portholes. This means that for present aircraft, interior designs are limited to long aisles, for all three classes, each with a limited number of possibilities to add extra comfort. Thus, the interior designer must attempt to improve features through a creative use of space and application of materials.

Special zones

Traditionally, airlines have adopted the concept of keeping the passengers in their seat and bringing an increasingly larger range of services to the individual. Now that the scope of these services is quite large, and passengers wish to select which ones to use, it may be necessary to consider the cost of this approach against the concept of providing a number of areas with special functions.

This concept was mentioned earlier in connection with communication and computer requirements for passengers but the concept can be expanded to a wide range of other individual requirements. These could consist of dividing spaces into specific zones such as those for seating, sleeping, hygiene and movement. Optional areas could also be for the disposition of business people with special features such as communication facilities. Other special areas could provide a nursery or games room for children, whilst appropriate facilities could also be provided for people of reduced mobility. Whilst account is now taken of special groups the concept of special areas might be able to enhance the attention which can be given, especially where the need is essential such as aged persons with special requirements, handicapped persons and others needing specialist medical help.

The viability of such concepts will depend on the passenger mix, in terms of such things as how much a telephone is used or the extent to which a computer or fax facilities are used. It might be that special areas for these activities would actually stimulate extra use by passengers who might not usually use them so much. There is also the possibility that the creation of special areas would allow seating arrangements to be configured differently.

The scope for using the cabin space will greatly increase with the advent, in the near future, of super large aircraft with more than 500 passengers. The resulting large spaces have parallels in other areas. These include the vast interiors of some ships, with their majestic walkways and grand staircases. In terms of using small spaces to the maximum the opposite also applies in the study of many high-tech areas, such as space station design or submarines.

Nevertheless, it is necessary to consider the different context for passenger aircraft and thus to recreate the concepts through adaptation and optimization of the new technologies. It is particularly important to avoid achieving sterile results through concentrating solely on 'cosmetic solutions'. Indeed, if some of the forthcoming aircraft have new shapes for their fuselages it provides new opportunities for the inside requirements. The advent of double curved hulls would certainly help to produce a 'straighter' wall in the cabin. The advent of the 'blended-wing' concept will be particularly interesting as the passenger spaces will branch off into the wing - producing a revolutionary experience for air travellers!

Personalising travel

The concepts involved in special zones will particularly lend themselves to 'personalising' areas by applying the appropriate ergonomics, colour harmonies and materials. The challenge of size certainly improves the scope for modularity of the elements. The scope for such variations will depend on the type of aircraft but as there will be a choice of aircraft speeds it will be necessary to separate the individual requirements. Business travel will call for minimum time between destinations together with a the best possible levels of service, whilst for social, or 'pleasure', travel relaxation and a pleasurable travel experience will be the key features. For the first requirement the choice of flight will be fairly simple but for the second the context and expectations will determine the choice of flight.

For all classes of travel, though, the interior design must focus on the flight as being a phase of activity for the passenger rather than just a means of transportation. In this sense the designer must think like an architect

Developing the Future Aviation System

who must assure the harmonious synthesis between the interior and the outside of building. Nobody likes a building, which is beautiful from outside but unpleasant to live or work in. Of course, buildings have been used for thousands of years by humans, most of whom consequently have their own opinions about them. However, aircraft are relatively new and until recently they possessed a certain mystique. As the new millennium approaches this mystique is fading as flight becomes part of the daily activity for many people.

Of course, the requirements for passenger safety will still remain, or even be enhanced, due to the new concept of interior design. Thus, the architecture of interior must put its methodology to the service of the passenger, travelling for business or leisure. It must take account of the passengers habits and desires whilst protecting the passengers from their fears and phobias and unexpected events.

Developing world-wide norms

As long-haul aviation implies a world-wide activity it is essential to develop a number of norms which can deal with this situation. For instance, there must be technical norms which can produce convergence on a world-wide basis, especially between American, European and Russian standards. This urgent need is essential in order to speed up innovative developments in the aeronautical sector.

Another area which needs attention is the ability of airports to load passengers onto aircraft which may have new configurations. For instance, the implementation of special zones might also allow embarkation of passengers by these zones, assuming the airports have provided sufficient loading positions. If the concept of special spaces in aircraft for specific groups is followed then it would certainly improve the pleasantness of the journey if those group passengers receive the same level of service at the airport.

It is also important to ask what is the basic level of bodily hygiene, which a passenger is entitled to? Passengers often request better hygiene so why can showers not be provided for long haul flights?

The complementary services

The design of an aircraft must consider other providers of services to an airline and the passengers. At airports this will include technical services (e.g. the firemen, the maintenance teams) and they will have a direct influence on the interior design of an aircraft which in turn will affect the passengers.

In the same way, an airline must remember that each traveller is an individual who has started from home and then passes through a whole series of travel stages in order to arrive at a final destination. Thus, as mentioned earlier, it is important to consider the context of a flight, with such matters which precede it and those which will follow it. For instance, having decided to travel somewhere, a passenger's first action will be to book a ticket. This can be made relatively simple or it can be made very complicated if the many options (e.g. types of communication required, special service needs, or nursery requirements, group seating needs) are offered.

On the other hand, there are variations which the passenger would recognise as being relevant to his, or her, needs. One very helpful feature would be to be offered different ways of handling passenger luggage. Some is suitable for the luggage hold, some is particularly needed during the flight whilst a third category, such as small items and fragile ones, could be placed in a shared luggage area in the cabin. This concept could be standardised by the use of mini containers, which would certainly be of help to the passenger and the aircraft operator.

Ergonomic principles

Features of the passenger cabin are generally based on the proportions of 'the average man', with little account usually taken of some important inherent differences due to the age, size, and weight of the passengers. Fortunately, some seats are now variable in width together with more variations for mechanical movement. However, the concept of identifying zones for different activities means that different types of seat might be employed and a more rational utilisation of seat spaces might liberate areas for other uses.

Other features, which differ strongly between people, are sounds and odours. These two aspects are, apparently, very different but they do both entail a certain degree of nuisance. Initially the level of impact might be tolerable but it becomes exacerbated with time - especially on a long flight. Consequently these real problems are mentioned here as more research is needed on them.

Unwanted noise is always a problem and so some advanced studies are in progress on this subject and include such possibilities as helmets to reduce the noise level.

Similarly, unwanted odours are just as annoying. Research in this area benefits from technical abilities to synthetically recreate numerous fragrances in some sectors although application of the fragrances is often limited due to the difficulties imposed by restricted spaces. Another problem is that the olfactory organs of individuals are different in sensitivity and sometimes non-objective.

However, careful application of aromacology is a technique, which if well mastered could relieve some particular pains and uneasiness or stress in general.

As for the other services, passenger expectations will drive the possibilities of choice which airline operators, and thus the aircraft manufacturers, must offer.

Conclusion

As mentioned earlier the 'tubular' fuselage creates a major constraint but there are plans for other shapes. Even with double story fuselages or catamaran styles the same problem of the 'tube' will remain. Extremely large aircraft, which may be veritable 'ferries' of the sky could still be tubular but the size will mean that the breadth would be significant so that activities across the aircraft become possible.

To foresee future modes of air travel it is necessary to presume that some of the present technological studies will come to fruition. Perhaps success will be achieved for ideas such as the blended wing (where some cabin space will be in the wing areas), or new generations of the dirigible will provide slow but spacious journeys. Perspectives will have to change dramatically - particularly if flight into geostationary orbit becomes feasible.

These ideas might be thought of as Utopian, but the aeronautical world continues to follow technological developments and so it is important to apply the same progress in the interior design of aircraft.

References

Air & Cosmos/ Aviation International, M 2756 Hors-série
Beauchamp J, (1994) *Avenir no.1- Tourisme 1994-2001*
Breand A. (09/97), Virtual aiplane Cabins materializing, *Revue Aerospatiale No. 141.*
Le Concorde, (06/92), *Cahier de sciences & vie, Hors série No.9,* .
Couroux P. (12/95), in *Sciences & avenir* (pp 88-89).
Déribéré M.(1993), *La couleur,* Presses Universitaires de France.
De Noblet J. (1988), le geste et le compas , in Aimery Somogi (ed.), *Design.*
Dollfus Ch. and Bouché H. (1942), 'Histoire de l'aéronautique', in *L'Illustration.*
Revue Aerospatiale No. 142 (10/97), A3XX A true Air Liner

Dubois Th. (07/97), Les Limousines du ciel (pp 46-48) in *Air & Cosmos/ Aviation International* No. 1621-22.
Edwards S. & the editors, 'Product design 2', *ID Magazine*.
Fillacier J. (1986), *La pratique de la couleur*, Dunod
Gallimard ed. (10/97) *Les dirigeables - L'épopée méconnue des Géants du ciel*
Huet S. (11/90), *Sciences & avenir No.525*, pp 46-48.
Itten J.(1977), *Art de la couleur*, Dessain & Tolra
Jacini M. (1994) in De Vecchi (ed.), *Cours d'architecture d'intérieur*.
Learmount D. (12-18/11/97), Passengers on the rack, *Flight International*.
Kobayashi S.(1990), Color Image Scale, *Kodansha International*.
Lesquin M. (10/97), The biggest Bizjet Available soon, *Revue Aerospatiale No. 142 - A319*.
Lenclos J.Ph. & D. (10/95), Les couleurs de l'Europe, in *Le Moniteur*.
Le Temps Apprivoise (1993) Zech P. Ed., *La couleur, Nature*.
Mefret J.M (31/10/97), Dossier sur les 25 ans d'Airbus, *Figaro Magazine*.
Nicolaou S. (1997), Les premiers dirigeables français - E.T.A.I & Musée de l'Air et de l'Espace, *Coll Envols no.3*.
Sarsfield K. (12-18/11/97), Insiders trading, *Flight International*.
Sweetman B. (06/95) *Air & Cosmos/ Aviation International No. 1523*, p 70.
The Art Institute Of Chicago(10/96), *Architecture & design for Commercial Aviation*, ed.Zukowsky J., Prestel.
The Creative Process Behind Product Design, *Industrial Design Workshop 2*, Meisei (ed.).
Zwimpfer (1985), *Couleur, optique et perception*, Dessain & Tolra

The author

Carole Favart-Andrieux lives in Paris, France, with her husband and two children. She qualified in 1981with a post-graduate degree in interior design from the Ecole Nationale Superieure, Arts Appliqués & Métiers d'Art, in Paris.

Besides being a part-time teacher in design schools she established her own company, Relief Design, in 1990. Her work includes interior design for aircraft and boats, product and graphic design as well as trend and colour consultancy.

The main customers are Brandt, Dassault Aviation, Dassault Falcon Service, Groupe Brandt, Jet Aviation, Lacoste Lunettes and Vedette.